POTATO

Production, Storage and Marketing

POTATO

Production, Storage and Marketing

by :

Narendra Singh, Ph. D. (Horticulture)
Scientist

Zakwan Ahmed, Ph. D. (Plant Breeding)
Former Director

Defence Research & Development Organization
Defence Institute of High Altitude Research, Leh-Ladakh (J&K)

SSPH

SATISH SERIAL PUBLISHING HOUSE

403, Express Tower, Commercial Complex, Azadpur, Delhi-110033 (India)
Phone : 011-27672852, Fax : 91-11-27672046
E-mail :info@satishserial.com, hkjain1975@yahoo.com
Website : www.satishserial.com

Published by :

SATISH SERIAL PUBLISHING HOUSE

403, Express Tower, Commercial Complex, Azadpur,
Delhi-110033 (INDIA) Phone : 011-47073040
E-mail : info@satishserial.com, hkjain1975@yahoo.com

Print Again : 2023

ISBN 81-89304-74-7

Composed, Designed & Printed in India

भारत सरकार, रक्षा मंत्रालय
रक्षा अनुसंधान तथा विकास संगठन
201, डी.आर.डी.ओ. भवन, नई दिल्ली–110011
Government of India, Ministry of Defence
Defence Research & Development Organisation
201, DRDO Bhawan, New Delhi-110011

डॉ० डब्ल्यू. सेल्वामूर्ति
विशिष्ट वैज्ञानिक एवं मुख्य नियंत्रक अनुसंधान तथा विकास
Dr. W. Selvamurthy
Distinguished Scientist & Chief Controller Research & Development

MESSAGE

I am pleased to learn that the scientists of Defence Institute of High Altitude Research (DIHAR), Defence Research and Development Organisation, Leh–Ladakh (J&K) are bringing out a book entitled **"Potato: Production, Storage and Marketing"**, which is based on the research carried out during more than a decade in the region.

The technologies generated at DIHAR have revolutionized potato production specially in most challenging hostile high altitude cold arid region. Potato is not just a vegetable but also a wholesome food and is being recognized as an important component of dietary intake to provide food and nutritional security.

The scientists have also developed terrain suited post harvest technology to ensure eco–friendly zero–energy based safe storage and round the year availability of locally produced potatoes for civil population and Armed Forces deployed in the region. The technology is also useful to prevent huge annual losses of farm produce.

I send my warm greetings to the authors on this occasion and hope that the book being brought out will be well received by all concerned.

Dated: 27 August, 2009 **(W. SELVAMURTHY)**

भाकृ अनुप
ICAR

Central Potato Research Institute
(Indian Council of Agricultural Research)
Shimla - 171 001, HP (India)
Fax: 0177-2624460 Telegram: Potatosearch, Shimla

CPRI

Dr. S.K. Pandey
Director

MESSAGE

The Leh -Ladakh region situated in the high hills lies in cold desert zone of north western Himalayas. The harsh climatic conditions in the region offer only limited period *i.e.*100-120 days for growing agricultural crops during summer season (May –September). Potato a short duration crop having highest dry matter production/area/time can play a pivotal role in raising farm income of the people and meeting their food and nutrition security. Predominance of local varieties and age-old farm practices are the major bottleneck for the low productivity in the region. Raising potato productivity under such difficult situations was a challenging task. The scientists had no option but to attempt a solution to the problem by developing technology on potato research relating to variety evaluation, developing package of practices, TPS and its utilization for potato production, various aspects of storage in the inhospitable and adverse climatic conditions of the region.

The painstaking efforts undertaken by the team of dedicated scientists from the Defence Institute of High Altitude Research, Defence Research & Development Organization, Leh-Ladakh for bringing out the book on **"Potato: Production, Storage and Marketing"** especially Dr. Narendra Singh deserve appreciation. The technology outlined in the book is likely to boost potato production in the region thereby providing round the year availability of ware and seed potato not only to the local population but also meeting the important demand of the defence personals employed in this sector.

The book is likely to serve as a benchmark for the future generation of the scientists, extension workers and more important to the farmers engaged in potato cultivation in cold desert regions of the country.

August, 2009 **(S.K. PANDEY)**

Govt. of India
Ministry of Defence
Defence Research &
Development Organisation
Defence Institute of High
Altitude Research
PIN : 901205, C/o 56 APO

Dr. Shashi Bala Singh
Director

FOREWORD

Cold arid region comprises of uneven land, barren mountains, arctic desert and extremely harsh climate which makes agricultural operations possible only during short period of summer season (May - September). To grow any crop in this limited period, specific cultural practices are of utmost importance. Potato is a short duration crop that produces more dry matter per unit area and time than any other major crops like wheat, maize and barley. Therefore, the problem of food and nutrition security can be overcome up to some extent by boosting potato production and it's round the year availability. With this aim the research was carried out on various aspects of the potato crop.

Various recommendations from the research trials carried out in high altitude cold arid region of Ladakh such as production technology for table and seed potato, varietal suitability, crop health management, true potato seed (TPS) technology and potato production in greenhouse were provided to the farmers which resulted in enhancement of productivity of the crop. Potato cannot be stored under natural conditions in cold arid climate therefore with increase in production, recurring temporary glut was being faced by the farmers in the recent past. To sort out this problem Defence Institute of High Altitude Research has developed a technology for long-term potato storage in specifically designed underground structures with CIPC treatment. This zero-energy based technology has showed immense potential for on farm storage of farmer's produce at low cost and making quality potato available during winter months.

I hope that the information generated by research trials will benefit farmers and organizations involved in potato based scientific and industrial strategies.

Dated: 30 July 2009

(SHASHI BALA SINGH)
DIRECTOR

PREFACE

Most of the research work in India pertaining to development of suitable potato varieties, standardization of package and practices, disease free seed production technique, insect – pest, diseases and their control measures, suitable varieties for processing/potato based industries, post harvest management, storage etc. have been carried out in the physiographic conditions of plains and lower hilly regions. But negligible practical research work has been carried out on this important crop for high altitude cold desert physiographic conditions.

The high altitude cold desert regions are situated in harsh remote areas at very high hills (more than 2700 meters) and they are separated out for 6–7 months in a year by surface transportation from rest of the world. Some of the regions/valleys cut off again from each other during severe winter months due to heavy snowfall. Large number of villages situated at very high altitudes depends only upon agriculture in the region. The areas have very short (June–August) agricultural season and its main dependency is on barley as staple food. Potato as a staple food crop can play an important role for food and nutritional security in these regions.

Considering the above constraints, research work was carried out on various aspects such as production technology for table and seed potato, TPS (true potato seed) technology, crop improvement, crop protection, crop sustainability, post–harvest handling and use of geothermal energy for storage during freezing winter and long term storage using Isopropyl 3-chlorocarbanilate (CIPC) for round the year availability of locally produced potato. The research work carried out on the above aspects for more than one decade is described in various chapters of this book. We hope that the compilation will be beneficial to researchers, academicians, policymakers, traders, growers, students, NGOs and potato based industries and all those who are associated with various missions to develop the hilly regions.

Indeed, we are very appreciative of the support and co-operation extended by Defence Research and Development Organisation (DRDO) for various R&D projects and feasibility study on potato crop. The authors are also grateful to Shri M. Natarajan, SA to RM, Dr. W. Selvamurthy, CC R & D (LS & HR) and Dr. RC Sawheny, Director, Dte of Life Science, DRDO HQ New Delhi for their guidance, blessings and encouragement. We wish to express our sincere thanks and gratitude to Dr. Brahma Singh (the then Director, Dte of Life Science, DRDO HQ, New Delhi) for emphasizing, guidance and encouragement of the research work on potato crop being among the fresh vegetables it supplied in highest quantity (about 40%) to the Armed forces deployed in the region.

Grateful thanks are cordially extended to Dr.S.K. Pandey, Director, CPRI, Shimla and his team of expert scientists for critically going through the manuscript and their valuable suggestions to improve this manuscript. The assistance provided by Dr. Bhopal Singh, Senior Scientist, Seed Production Division, IARI,New Delhi, Dr. Gunjeet Kumar, Scientist, NBPGR, New Delhi and Dr. S. R. Kushwah, TO, Plant Breeding Division, IARI, New Delhi in statistical analysis of the data for preparation of this manuscript are thankfully acknowledged. We acknowledge with thanks to the farmers of Nang village for adopting the quality seed production techniques and production of quality seed on pilot scale, which has significant impact on boosting the potato crop in the region. We are deeply indebted to late Shri Rigzen Namgyal Kalon (the then Chairmen, Farmers' Cooperative Marketing Society, Leh-Ladakh) who assisted in potato seed village programme to make it successful venture. A deep sense of gratitude goes to Ladakh Autonomous Hill Development Council, Leh-Ladakh for valuable suggestions based on the farmers' requirement for the direction of the research programme. To conduct the FLD trials, assistance provided by State Agriculture and Horticulture Departments, Leh and SKUAST Research Station, Leh are thankfully acknowledged. Our sincere thanks are due to our colleagues for the help rendered in various ways in the completion of this book. We also wish to express our appreciation to the employees of the Institute who helped in the field trials conducted during the period. The authors would like to extend their sincere thanks to Shri Hemant Jain, M/S Satish Serial Publishing House, Delhi for his quality and timely printing of this manuscript.

Authors

CONTENTS

CHAPTER **1**

Introduction

(1)

(2)

(3)

(4)

Plate 1: Different type rocks (1-3) and cultivable fields (4) of cold desert region

In India high altitude cold desert regions under the Trans-Himalayan zone are confined mainly to Ladakh in Jammu & Kashmir (J&K) and Lahul & Spiti in Himachal Pradesh. In J&K cold desert lies between 32°–36° N latitude and 75°–80° E longitude and covers approximately 68,321 Sq Km area whereas in Himachal Pradesh, it lies between 31°-32° N and 76°-78° E covering an area of 6,488 Sq Km. Small pockets in Garhwal (Niti and Mana) beyond Badrinath and Nelang region in Uttarkashi district of Uttrakhand state have similar climatic conditions and terrain (Chaurasia and Singh, 1996).

The climate in cold desert regions is unstable, fluctuating and unpredictable that makes the ecosystem most fragile. Due to very high altitude and little to no rainfall, the climate is cold and dry. The cold desert regions of India have been divided into seven valleys namely Indus (Leh), Nubra, Changthang, Zanskar, Suru, Lahaul and Spiti. The physical features of the valleys are characterized as coarse, porous and immature sandy soils prone to wind and water erosion, heavy influx of infra red and ultraviolet radiations, low air density–reduced oxygen level, low concentration of atmospheric carbon dioxide, long (5-7 months) freezing winters, short cropping season, large variation in seasonal temperature (40° C to minus 40° C), large diurnal variations in daily temperatures (0° C to 35° C) during cropping season with winds speed of 40–60 km per hour mainly in afternoon hours. Precipitation occurs mostly in the form of snow during winters. Irrigation water is available mostly from snow melts. Low relative humidity remains during crop growing season while, during winter soil moisture remains frozen. Therefore, the high altitude cold desert regions pose a great challenge in raising crops. The undulating land, isolated fields, inhospitable climate, availability of limited inputs, ethical and religious fanaticism are some of the other constraints in the region.

High altitude cold desert region faces extremely harsh climate with winter temperature as low as -30° C at Leh and -50° C at Drass and the summer temperature goes upto 35° C. January is the coldest month where maximum day temperature also remains subzero. Due to heavy snowfall, humidity increases from the month of December-February. July remains the hottest month with the lowest humidity. The temperature remains below 0° C for over 7 months in a year (Fig. 1.1).

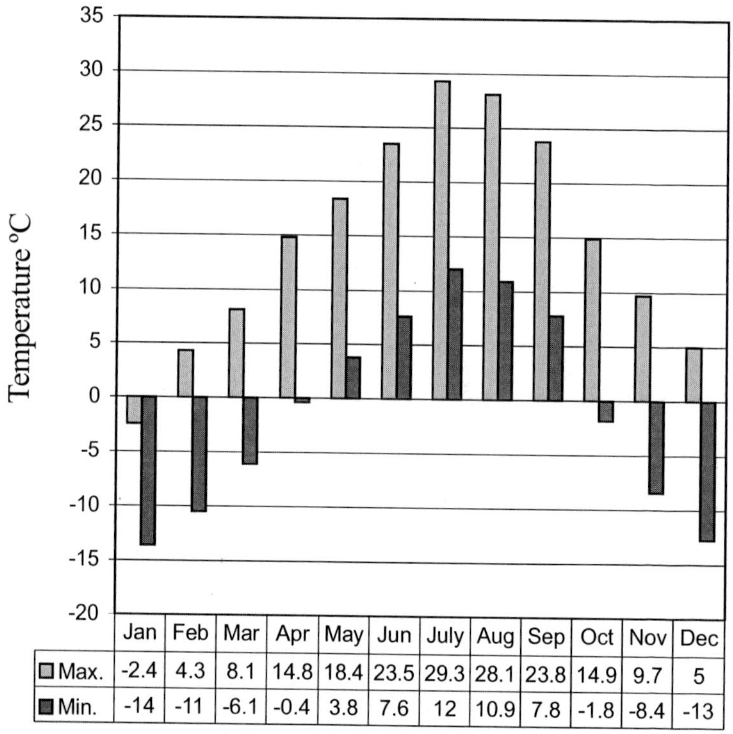

	Jan	Feb	Mar	Apr	May	Jun	July	Aug	Sep	Oct	Nov	Dec
Max.	-2.4	4.3	8.1	14.8	18.4	23.5	29.3	28.1	23.8	14.9	9.7	5
Min.	-14	-11	-6.1	-0.4	3.8	7.6	12	10.9	7.8	-1.8	-8.4	-13

Fig. 1.1: Average temperature (°C) of Leh (Altitude 3450 meters)

The remote areas have a direct relationship with food and nutritional insecurity leading to poor health and loss of human potential. Providing food to the troops poses the greatest challenge because of their deployment on remote frontier locations and timing of operations in these areas. This has been aptly summarized by Frederick as "An army frights on its belly" and by Napoleon as "An army marches on its stomach". These sayings are equally important to all the human beings. All over the world, armed forces are supplied food according to the ration scales designed by nutritionist so as to suit food habit of the troops, place of deployment, climate and availability. The basic recommended quantity specially of vegetables for Indian soldiers in high altitude per head per day is 140 g potato, 170 g fresh vegetables (included more than 20 types of vegetables) and 60 g onion. Among the vegetables supplied to the troops, potato account for 38%, thus it needs more attention to study its production, storage and supply in this region.

Potato is a short duration crop producing more dry matter per unit area and time than cereal crops like maize and wheat. It is a versatile food and certainly the most valued non-cereal food crop. Potato is also a wholesome food. In terms of production, potato ranks fourth after rice, wheat and maize in the world. It is a rich source of high quality proteins, minerals and vitamins and superior dietary fibers. The potato proteins are rich in essential amino acid-lysine (Ezekiel *et al.*, 2003). Potatoes can be cooked in many ways viz. boiled, fried, roasted, baked or steamed. It can also be processed into a number of products like chips, French fries, flakes etc.

To maintain the regular supply of potatoes to the civil population of the region and the troops at our frontiers like Siachen, Turtuk, Kargil, Batalic, Drass in Ladakh sector of Jammu & Kashmir and along China border is not an easy task as the region remains land locked for over seven months (November to May) in a year due to heavy snowfall which closes the passes. Regular supply of fresh potato is also very difficult even when the roads open (June-October). Moreover long distance transportation add cost and led to huge losses. Potato air lifted from other parts of the country normally cost about Rs. 70 per kg by the time it reaches to the forward areas.

Lot of research work has been done to develop suitable agro-techniques to boost the production of potato and its storage and utilization in physiographic conditions of plains and low hill regions. On the other hand little work has been carried out on potato cultivation is the high altitude cold arid regions, whose climatic and geographic differentiation segregates the cold desert region from rest of the country. Hence, techniques for potato production, post harvest handling, storage, processing, utilization being employed else where, do not hold good as such for the regions. Although, long photoperiod, negligible rainfall and low incidence of major diseases and pests favour high productivity of potato, yet high altitude regions have certain bottlenecks for high production and productivity due to lack of scientific knowledge on various aspects of the crop.

In a survey it was observed that the farmers of the region had been growing potato since long time (30–40 years). Around the Leh and Kargil city, and Karu and Nubra regions, farmers grow potato on commercial scale and supply the produce to local market and Army. But in few far–flung villages, the potato is mostly grown for their own consumption. Two types of local varieties i.e. andigena and tuberosum group were introduced long back from Shrinagar. Majority of the farmers uses seed from the table crop for next crop. Some farmers exchange the potato seed with the farmers of other locations. State Agriculture Department also provides seed to the farmers on subsidized rates. Potatoes being a vegetatively propagated crop, many diseases are transmitted through tubers from one generation to another, which deteriorates the quality of seed tubers.

Cold desert regions have peculiar climatic and geographical conditions and the regions are situated in isolated places. Therefore, suitable cultural practices developed for one place do not hold good for the entire region. To get optimum yield, the potato crop needs terrain suited cultural practices and refinement in the potential traditional practices of potato production in the region. Therefore, due to inadequate knowledge the farmers are growing potato crop in a traditional way with out following proper agronomic-practices.

Only single cropping season of 65 to 150 days is available in the cold desert region, for potato crop (Table 1.1). Therefore, the crop cultivated during summer season is harvested in September needs to be stored for continuous availability in the long freezing winter reason. The farmers store their potato in underground pits (1.2–2.1 meter deep) at their field. The length–width of the pit depends upon the quantity to be stored. The pit is filled up in the month of October. They put 45–60 cm soil on the potato tubers in the pit. However, in these pits there is heavy weight loss and spoilage due to sprouting of the tubers. As the tubers are direct in contact with soil, the peel and its colour is badly affected after removing of the tubers from the pit. Due to rough appearance and texture of the tubers, the farmers get low price in the market. Generally farmers store the potato upto March in these pits. However, cropping season is normally starts from May onwards depending upon the altitude. Therefore, further storage of these tubers becomes difficult from March to May. Shriveling effect was observed almost on all the tubers, which may be due to water loss and heavy sprouting due to relatively higher temperature & low humidity (Fig. 1.1 & 1.2). Mostly farmers use potato tubers as seed after removing the sprouts from the tubers, which results in low yield. Potato is used in the name of fresh vegetable during long frozen winter in cold desert regions and generally it is cooked alone or with spinach.

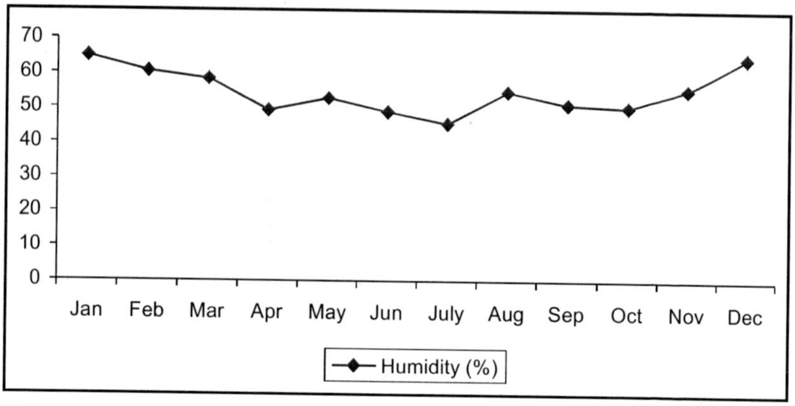

Fig.1.2. Relative humidity (%) in different months in Leh

Table 1.1: Potato crop season at different altitudes

S.No.	Altitude(m)	Duration of the agriculture season	Month
1.	2700-3000	May –September	5
2.	3000-3500	June – September	4
3.	3500-4000	Mid June – August	2.5

With successful cultivation of potato in these areas, the Armed Forces will get fresh and good quality potato comparatively at cheaper rate, and save lot of Government money and efforts to import of this commodity from other parts of the country. As such a detail study was carried out on various aspects of the crop. In the first step of the study, a detail field survey was conducted to know the traditional methods followed by the farmers for production of potato in high altitude cold desert regions. Despite a major food crop of the region, it was revealed from the survey that farmers were facing several technical problems. After the field survey, a programme on systematic research work on various priority aspects of this important crop was designed with emphasis on varietal performance trials. The breeders' seed of potato variety Kufri Chandramukhi procured from the Central Potato Research

Institute, Shimla was given to the farmers and it was further multiplied through concept of seed village and provided to the farmers through State Agriculture Department. Simultaneously over 30 research trials were conducted on production technology for table and seed potato, true potato seed (TPS) technology, post harvest handling, storage, crop improvement, crop protection etc. To get more reliable results, these trials were repeated for another two years.

The out come from the trials interms of trechnical information was provided to the farmers through training programmes and demonstration trials on their fields. As an immediate impact of the programme, the production of the crop was increased. But the increased production led to storage problem. To over come this, geothermal energy based store was designed for storage of potato during freezing winter. Trials were also conducted to ensure round the year availability of locally produced potato through long term storage with Isopropyl 3-chlorocarbanilate (CIPC) treatment The experimental trials were laid out in randomized block design and split plot design in different number of replications (Gomez & Gomez, 1984) and the data were analyzed using the software SPAR-I (IASRI, New Delhi), M STATE-C (Michigan State University, USA) and Microsoft Excel (MS Office). The findings of over thirty trials have been described in details in the different chapters of this compilation.

❑❑❑

CHAPTER 2

Evaluation of Varieties for Performance

Tubers of Sindhu Red during harvesting

Tubers of Sindhu Red after curing

Plate 2

Evaluation of Potato Cultivars for Yield Performance

In sandy to sandy loam soils of high altitude cold desert regions, only a summer crop (May–September) is grown. Therefore, higher production in the season is a priority aspect and can be achieved by growing suitable variety under scientific management practices. Commercial attributes of potato varieties are influenced by environmental factors to a considerable extent (Purohit *et al.*, 1970), therefore, suitable variety is the most important factor for potato production in a particular climatic conditions prevailing during cropping season. An investigation was therefore, carried out to evaluate production potential of some potato varieties. A field trial was conducted during 1997-1998 at Defence Institute of High Altitude Research, Research Station Ranbirpur Thickshey, Leh at altitude 3600 meters above mean see level.

The experiment was conducted in randomized block design with three replications with eight varieties *i.e.* Kufri Badshah, Kufri Chandramukhi, Kufri Jyoti, Local–1, Kufri Bahar, Kufri Jawahar, Kufri Ashoka and Local–2. Uniform potato seed tubers of all the varieties were planted at 60x20 cm in plots measuring 2.40 x 2.40 m in first week of may (6th May). 50 kg N, 80 kg P_2O_5 and 60 kg K_2O per hectare were applied at the time of planting while 50 kg N per hectare was given after 50 days of planting at proper moisture level followed by earthing up. Field was kept weed free by two weedings and six irrigations were provided as and when required. The crop was dehaulmed at 125 days after planting and harvested in last week of September. Tuber yield, average tuber weight and percentage of marketable tuber yield (> 20g tubers) were recorded.

The data (Table 2.1) indicated significant difference among the varieties for total tuber yield. Variety Kufri Chandramukhi recorded the highest tuber yield (388 q/ha) followed by Kufri Jyoti (310 q/ha), Kufri Badshah (286 q/ha), Kufri Jawahar

(280 q/ha), Kufri Ashoka (279 q/ha), Kufri Bahar (252 q/ha), Local–2 (244 q/ha) and Local–1 (201 q/ha), respectively. Kufri Chandramukhi had shown the significant superiority which may be due to suitability of its genetic make up for high altitude cold desert climatic conditions (Kadian *et al.*, 2003).

Table 2.1: Tuber yield, tuber weight and marketable yield performance of potato varieties (pooled over 1998-99)

Treatment	Tuber yield (q/ha)	Average tuber weight (g)	Marketable yield (%)
Variety			
Kufri Badshah	286	83	94.1
Kufri Chandramukhi	388	89.1	95.3
Kufri Jyoti	310	80.5	94.7
Local - 1	201	70.9	91.6
Kufri Bahar	253	65.9	93.6
Kufri Jawahar	280	76.5	92.8
Kufri Ashoka	279	80.7	94.0
Local - 2	244	74.5	90.5
CD (0.05)	87.9	20.4	3.0

However, higher tuber weight and marketable yield percentage non-significant were recorded with Kufri Chandramukhi but statistically non-significant differences were observed among the other varieties. The results revealed that Kufri Chandramukhi can successfully be grown with higher production and % of marketable yield in high altitude cold desert climatic conditions.

Evaluation of Various Advance Potato Cultures for Yield Performance

One red peel culture (Sindhu Red) was developed at Defence Insitute of High Altitude Research, Leh and in view of importance of potato crop in high altitude cold desert regions, as per a mandatory requirement, work was undertaken to evaluate various promising advance potato cultures to find out suitable high yielding variety for the region. Some advance potato cultures provided by CPRI Campus, Modipuram were tested under Ladakh conditions. Kufri Chandramukhi was included as local check. One traditionally grown culture was also included in the trial.

Eight potato cultures were evaluated at the Research Farm of Defence Insitute of High Altitude Research, Leh-Ladakh during 2002 and 2003 in summer season (May–September). Five potato cultures *i.e.* Kufri Sultan, Kufri Giriraj, Kufri Pukhraj, Kufri Jawahar and Kufri Jyoti were procured from CPRI Campus, Modipuram. The tubers of Kufri Chandramukhi (local check), Sindhu Red (developed at DIHAR, Leh) and Local–1 (traditionally grown) were selected from previous crop and stored during severe winter with the storage method developed by the Institute. Potato tubers (40–50 g size) of all the cultures were planted in complete randomized block design with three replications in plot size of 2.50 x 2.40 m at 50 x 20 cm spacing. Normal cultural practices were followed as described in earlier trial. Data were recorded on plant height at haulm cutting stage (125 DAP) and average tuber weight, total tuber yield and percent marketable tuber yield at harvest.

The pooled data from Table 2.2 showed significant differences among the cultures for the four characters studied. Significantly higher plant height was recorded with Kufri Sultan followed by Kufri Pukhraj and was minimum in Kufri Jawahar. The Kufri Sultan variety has the character to grow fast in the climatic conditions of cold arid region thereby having high plant height.

The perusal of data (Table 2.2) showed that average tuber weight differed significantly. Kufri Chandramukhi (76.03 g) produced tubers with higher average tuber weight followed by Sindhu Red (65.17 g). Local–1 (48.70 g) had the lowest average tuber weight. Kufri Chandramukhi had the maximum average tuber weight, which may be due to less number of tubers per plant in comparison to other cultivars, which resulted in more foodstuff accumulation in lower number of tubers.

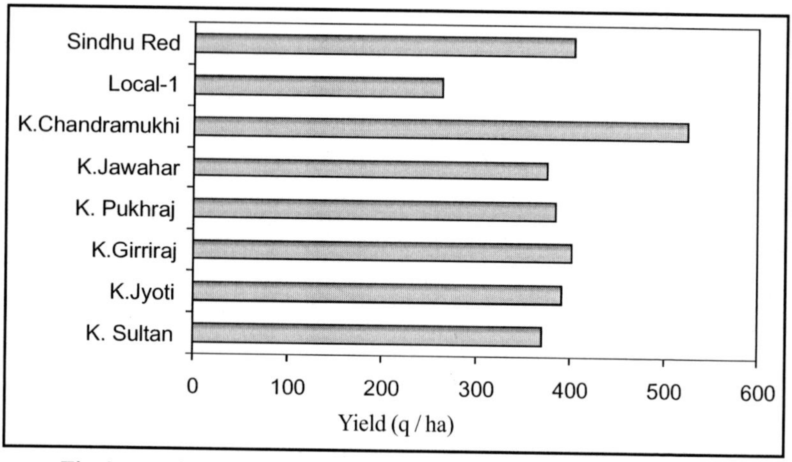

Fig.2.1. Tuber yield performance of different potato cultivars

Kufri Chandramukhi produced maximum tuber yield (524 q/ha) followed by Sindhu Red (403 q/ha) and Local–1 (264 q/ha) recorded the lowest tuber yield (Fig. 2.1). Total tuber yield of Sindhu Red and Kufri Jyoti was at par. Kufri Chandramukhi yielded higher tuber yield because of more number of tubers of large size and it was well acclimatized in the climatic conditions.

Table 2.2: Average plant height, per tuber weight, total tuber yield and marketable yield of different potato cultivars (pooled over 2002-03)

Treatment	Plant height (cm)	Av. tuber weight (g)	Tuber yield (q/ha)	Marketable yield (%)
Variety				
Kufri Sultan	90.4	60.83	370	92.3
Kufri Jyoti	60.2	60.10	390	92.6
Kufri Giriraj	56.9	61.13	402	96.4
Kufri Pukhraj	82.6	54.03	383	90.8
Kufri Jawahar	70.2	56.20	375	91.3
Kufri Chandramukhi	66.2	76.03	524	95.9
Local – 1	73.3	48. 70	264	87.3
Sindhu Red	73.4	65.17	403	93.6
CD (0.05)	3.0	1.43	34.6	1.4

Percent marketable tuber yield differed significantly with maximum percent marketable tuber yield recorded with Kufri Giriraj was at par with Kufri Chandramukhi and both the cultures were significantly superior over others. Local-1 gave lowest % marketable tuber yield. The higher percent marketable yield of Kufri Giriraj and Kufri Chandramukhi might be due to the number of large size tubers. The results of the experiment indicate that Kufri Chandramukhi and Kufri Giriraj are the suitable to grow in high altitude cold desert climatic conditions.

❑❑❑

CHAPTER **3**

Seed & Its Multiplication Case Study

Seed multiplication of potato cultivar Kufri Chandramukhi at farmers' field in Nang village

Potato seed village (Nang) concept appreciated by H'ble Dr. APJ Abdul Klam

Plate 3: Seed multiplication of potato variety Kufri Chandramukhi at farmer's field in different villages

Case Study on Quality Seed Production

Vegetative mode of propagation makes the seed potatoes a very important component of potato cultivation. Potato is a good host to many pests including viruses, phytoplasmas, bacteria, fungi and insects. Once infected by a pathogen, particularly virus(es), seed stocks progressively show decline in the tuber yield due to progressive increase in the infection level. Seed potatoes are a costly commodity and constitute a significant part of the cost of cultivation. Economic cultivation of potato warrants regular supply of healthy seed potatoes. Cold desert region of Ladakh (J&K) comprises areas, which are suitable for potato cultivation. Congenial climate of this region ensures high tuber yields provided healthy seed stocks are planted. However, these areas are very remote and transportation of bulky material like seed potatoes is very expensive and is almost out of the reach of the farmers. Therefore, the farmers use potato seed stocks, which are very old and are riddled with potato viruses. As a result, the full potential yield of potato which otherwise is very much achievable in the region is not harvested. Further potato is also a very important food crop for the region. Considering the significance of potato crop in the region, efforts were made to boost the production by providing quality seed to the farmers under a scheme named "seed-village concept" to make them self sufficient in healthy seed potatoes. The scheme involved selection of villages where vector pressure was negligible and healthy seed could be multiplied without degeneration. To start with village Nang (3750 m altitude) about 30 Km from Leh city was chosen due to low vector pressure (Garg *et al.*, 2003) for multiplication of seed potatoes. Breeder seed of potato cv. Kufri Chandramukhi from Central Potato Research Station, Jalandhar was airlifted from Chandigarh to Leh in April 1996 for multiplication as foundation seed through State Agriculture Department at this village. From the foundation seed produced at Nang village, three hundred quintals was procured by State Agriculture Department, Leh and it was further multiplied into certified seed at other suitable villages namely Karu, Leh, Khalsi and Diskit of Leh District. To enhance the supply of healthy seed potatoes, another 600 quintals of foundation–II seed was also provided from Nang village to the Agriculture Department as well as farmers for further multiplication as healthy seed. These efforts helped in replacement of the old home grown seed of local varieties in the villages with healthy seed of variety Kufri Chandermukhi. Selected seed growers were taught the improved agro–practices through trainings and field demonstrations. The availability of quality seed to the farmers resulted in a significant boost to potato production in the region. Efforts of DIHAR, in this regard resulted in increased meeting of potato requirement of the Army deployed in the region from 18% during 1998 to 36% in 2001 and further to 60% during 2005 (Fig. 3.1). This helped Army to have quality produce locally without incurring huge expenditure on air freight and local farmers more income from higher productivity of healthy seed.

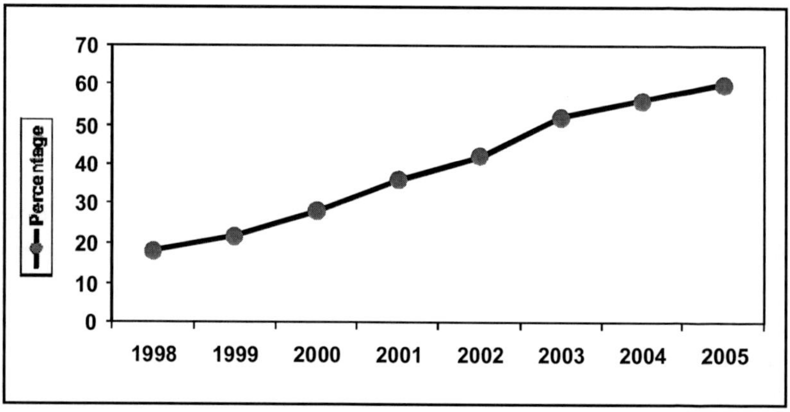

Fig. 3.1. Impact of quality seed tubers on supply of potato (% of total requirement) to army

To further boost production of potato and its availability to the Armed Forces, a multi–aspect research programme was carried out simultaneously. Under the programme, various experiments were conducted on evaluation of varieties, find out terrain suited agro–techniques for production of table and seed potatoes, exploitation of True Potato Seed (TPS) as alternate technology for quality seed availability in the region, development of suitable variety, potato production in greenhouse and mulching effect on productivity. To maintain continuous supply of potato for greater period of the year, intensive studies were carried out on storage of potato under cold desert climatic conditions. The significant findings of the studies were provided to the farmers from time to time through Co–operative Marketing Society, State Agriculture Department, Ladakh Autonomous Hill Development Council, Leh & Kargil and Shere Kashmir University of Agriculture and Technology (SKUAT), Research Station, Leh, training programmes, visits and front line demonstrations at farmers fields.

Therefore, 300 quintals of breeder seed of variety Kufri Chandramukhi was again procured from CPRI, Shimla and distributed through the Farmer's Co–Operative Marketing Society and State Agriculture Department to the farmers of different villages for multiplication into foundation and certified seed. The programme is being carried out by the farmer's society and the State Agriculture Department with technical consultation of DIHAR.

To include more number of villages in the seed village concept for multiplication of quality seed in the region, a Joint team of scientists from CPRI, Shimla and DIHAR surveyed the area for vector pressure and health status of potato crop to find out localities suitable for healthy seed multiplication. Aphid counts were taken in the crop by adopting the standard technique of 100 leaf count. Data recorded from different locations is given in (Table 3.1) perusal of which revealed that *M. persicae* population was almost nil in Saboo and Nang localities while it was higher in and around DIHAR, Ranbirpur, Thiksey, Phyang and Nimmu localities. Due to low vector pressure, Saboo, Nang and some localities above 3600 m altitude were recommended for multiplication of breeder seed stocks.

Table 3.1: Aphid population and viral diseases incidence at different locations in Leh (Ladakh)

Location / Village	Altitude (AMSL)	Aphid count*	Remarks	Virus detected
FRL Farm, Leh	12,500	193	Crop planted during end March and ready for harvest in July	NT
Saboo (Upper part)	13,000	Nil	Overall good crop; 15% plants showed virus infection	PVX, PVA
Saboo (Lower part)	12,600	6	Majority of plant showed virus infection	PVA, PVY, PLRV
Ranbirpur	11,700	21	-do-	PVA, PVX, PVY, PLRV
Nang	13,200	2	Overall good crop; 20% plants showed virus infection	PVA, PVS
Shankar Shubhi	11,700	40	Crop showed mosaics and leaf roll	PVX, PVA, PVS, PVY, PLRV
Nimmu	11,500	2	60% plants showed virus infection	PVA, PVY, PLRV
Phyang	11,800	5	About 25% plants showed virus infection	PVX, PVS, PVA

*Number of aphids per 100 compounds leaves taken at random.

As expected, the crop health at upper Saboo and Nang, where vector level was very low, was comparatively better than other locations. The poor health of the crop in majority of the areas was mainly due to use of the same stock year after year. Analysis of samples using immuno electron microscopy at CPRI, Shimla revealed the saturation of stock with aphid transmitted viruses *i.e.* PVA, PVY and PLRV, which reduced tuber yield drastically (Khurana, 1999; Garg *et al.*, 2003).

❑❑❑

CHAPTER **4**

Scope of True Potato Seed (TPS) Technology

(1) Planting of TPS parental lines

(2) Crossing block

(3) Profuse flowering

Plate 4: Hybrid TPS Production (I)

Potato is normally grown vegetatively through seed tubers. In many developing countries potato cultivation is hampered due to non-availability of quality seed potatoes. Seed cost alone account for more than 40% of the total cost of cultivation. The repeated use of same seed stock, potatoes accumulation of degenerative viruses and other diseases in seed stock. None of the potato viruses reported in India is transmitted through TPS. Transportation of seed potatoes to remote areas like Ladakh is rather difficult. TPS as a low cost propagule not require expensive storage and transportation, has better scope in such areas than the traditional tuber seed system. Moreover, air lifting of the seed potatoes to Ladakh from northern plains further adds to the cost of seed. In addition the seed potatoes harvested in Feb/Mar in plains have dormancy and are not physiologically suitable for planting in April/May until these are treated to break the dormancy. In view of the above constraints with tuber seed system, true potato seed (TPS) stands good choice for potato cultivation in remote areas of Ladakh. TPS offers following advantages over tuber seed system:

i. TPS is a low-cost planting material. Only 150 g TPS is needed for growing transplanted potato crop in one hectare as compared to nearly 2.5 tonnes of seed tubers required to plant the same area.

ii. By using TPS, the seed potatoes can be used for consumption.

iii. Problems associated with tuber-transmitted diseases especially viruses are minimized through TPS.

iv. TPS derived seedling tubers cost approximately one-tenth the cost of quality seed tubers needed for planting one hectare.

v. TPS is convenient and inexpensive to store for several years under ambient conditions by maintaining low seed moisture content (< 5%) during storage.

vi. Easy transportation, due to the lesser seed requirement.

vii. TPS provides an opportunity to fit potato into different cropping systems as against tuber seed of correct physiological age can not be available to the farmers year round.

viii. TPS would also help in expanding potato cultivation into farming areas previously unable to produce potatoes due to lack of good quality seed tubers.

ix. Multiplication of healthy/quality planting material through TPS is much faster than traditional process of certified seed production.

TPS technology as an alternative method for potato production has two main components:

1. Production of true potato seed (TPS)
2. Utilization of TPS for potato production

True potato seed (TPS) known as botanical seed of potato results from sexual reproduction and is formed inside the fruits called berries. Through TPS ware potatoes can be produced by following two ways:

1. Through transplanted seedlings
2. Through seedling tubers

In first method seedlings are raised as in tomato crop in nursery bed by sowing true potato seed and are transplanted in field after attaining 10–12 cm height. In second method, in first season seedling tubers are produced in nursery bed and these are stored until next planting season.

❏❏❏

4.1
Hybrid TPS Production

Harvesting of male flowers

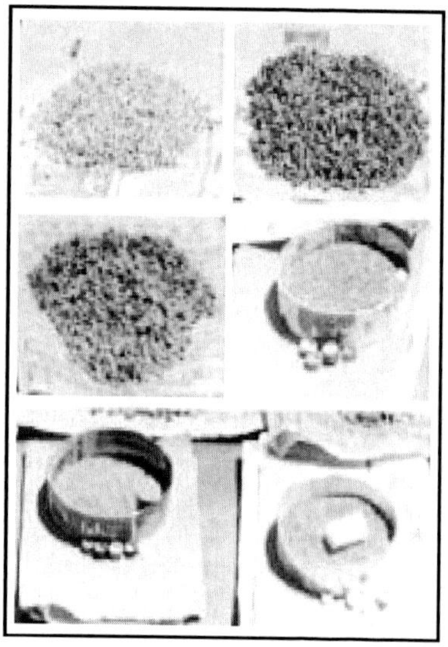

Pollen extraction from harvested flowers

Pollination of individual flower

Berry setting in female line

Plate 5: Hybrid TPS Production (II)

Performance of Different TPS Families for Hybrid TPS Production

Profuse flowering in the parental lines is essential for TPS production. Potato being a temperate crop needs cool climate and longer photoperiod for flowering to occur. In high altitudes, potato is grown during long summer days and the plants produce flowers under natural condition. Several characters are important for commercial production of hybrid TPS and its utilization for growing ware potato crop. These include high berry and seed set in the parents which facilitates production of more seeds per plant, seed germination, seedlings survival on transplantation and uniformity of produce.

Three female lines *i.e.* MF–I, MF–II and TPS–7 and two male lines namely TPS–13 and TPS-67 were used for hybrid TPS production at the DIHAR research farm, Leh. The female parents were planted on second week of May in 1996 and 1997 in randomized block design in four replications with the spacing of 50x20 cm (Plate 5). Male parents were planted one week before female parent in separate block maintaining of planting ratio of 4:1 (Female: Male). Hybridization, techniques were followed as describe in extension bulletin No. 30 of CPRI Shimla. Pollination was done twice (morning 8–10 AM & evening 4–5 PM) a day. Pollination was carried out up to first week of August. No thinning of stems, bunch or flowers was done in female parent and all the flowers in each bunch were pollinated. Support to the stems and cover to the bunches of berries were also not provided. All barries were collected in small gunny bags and the bags were kept for 10 days in polyhouse to give suitable temperature to soften the barries as ambient temperature is low at harvest in October (Fig. 1.1). Before storage, very small berries (<6g) were separated and seed was not extracted from them. After 10 days, seed was extracted with the help of hand Juicer machine. The juice and seed extracted from the barries were kept in greenhouse conditions for fermentation of the pulp. TPS was separated by washing and decanting 4–5 times. After washing, the seed was dried in shade. It was also exposed to warm sun to reduce seed moisture be fore packing for two storage.

The data (Table 4.1.1) showed that the female parents of the TPS families produced significantly varying number of flower bunch per plant. Higher number of flower bunch per plant was recorded with female parent MF–I (8.7 & 8.9) in both the blocks (MF–I x TPS–67 & MF–I x TPS–13, respectively). Same type female parents in both the blocks of the families were produced at par number of flower bunch per plant. Significantly maximum number of bunch per plant were produced with female parent MF–I possibly due to better genetic constitution and adaptability. Better genetic constitution of the female parent MF–I for more duration of flowering was also reported by Vaghela *et al.* (2007). However, more number of flower per bunch was recorded with female parent MF–II (11.7) but the difference among the families was non-significant. Maximum number of berries per bunch was recorded with TPS family MF–I x TPS–13 (5.6) followed by MF–II x TPS–67 (4.5). TPS family MF–I x TPS–67 (2.6) produced the lowest number of berries per bunch. Maximum number of berries per bunch in combination of MF–I x TPS–13 may be due to better combining ability and the genetic potential for adaptation in wider range of climatic conditions of the family. The similar results were also reported by (Kadian *et al.*, 2003).

Table 4.1.1. Performance of different TPS families in terms of number of bunch/ plant, number of flowers/ bunch and number of berries/ bunch (pooled over 1996-97).

Treatment	Number of bunch per plant	Number of flowers per bunch	Number of berries per bunch
TPS Strain Families			
MF– I x TPS–67	8.7	10.0	2.6
MF–II x TPS–67	7.9	10.6	4.5
TPS–7 x TPS–67	7.1	11.3	4.2
MF– I x TPS–13	8.9	9.6	5.6
MF–II x TPS–13	7.7	11.7	3.4
TPS–7 x TPS–13	7.0	10.9	3.6
CD (0.05)	0.3	1.2	0.5

Data in Table 4.1.2 revealed significant difference for number of berries per plant among the families. Significantly higher number of berries per plant was recorded in cross MF–I x TPS–13 (36.3) than all other families. Maximum number of berries per plant in MF–I x TPS–13 was perhaps due to higher rate of berry formation after pollination. The results of berry weight indicated that different families significantly influenced green berry weight. The cross MF–II x TPS–67 (10.0 g) produced bigger berries followed by MF–I x TPS–67 (8.7 g). The significant superiority of MF–II x TPS–67 might be due to the favorable genetical constitution of the parents. Total weight of green berries per plant also differed among the TPS families. Maximum weight of green berries per plant was recorded in MF–I x TPS–13 (662 g) which was significantly higher from other TPS families. More weight of green berries per plant in MF–I x TPS–13 may be due to more number of flowering bunches and number of berries per plant.

Table 4.1.2: Number of berries, average weight of green berry and weight of green berries per plant in different TPS families (pooled over 1996-97)

Treatment	Number of berries per plant	Average weight of green berry (g)	Weight of green berries per plant (g)
TPS Families			
MF– I x TPS–67	22.5	8.7	354.8
MF–II x TPS–67	28.2	10.0	471.2
TPS–7 x TPS–67	30.6	8.1	403.6
MF– I x TPS–13	36.3	7.9	662.3
MF–II x TPS–13	25.8	7.4	329.2
TPS–7 x TPS–13	25.2	7.9	303.9
CD (0.05)	1.6	0.4	23.1

The data presented in Table 4.1.3 indicated (323.0) significantly higher number of TPS per berry in MF–I x TPS–13 than the other TPS families, which may be attributed to suitability of the combination for maximum seed setting in the berry. Pandey and Gupta (1995) have also reported large variation in number of seeds per berry in different TPS families. However, maximum weight of TPS per berry was recorded in MF–II x TPS–67 (0.051g). Weight of TPS per plant significantly differed among TPS families. Higher seed weight per plant was recorded in MF–IxTPS–13 (2.37 g), which has shown superiority over other TPS families. The higher TPS production in MF–I x TPS–13 was mainly attributed to more number of berries per plant as compared to other cross combinations. The results are in accordance with earlier finding of Upadhya *et al.,* (1984).

Table 4.1.3: Effect of different TPS families on seed number and weight per berry, per plant and yield (pooled over 1996-97)

Treatment	Number of TPS per berry	Weight of TPS per berry (g)	Weight of TPS per plant (g)	Yield of TPS (kg/ha)
TPS Families				
MF– I x TPS–67	288.2	0.048	1.063	42.7
MF–II x TPS–67	240.1	0.051	1.80	74.41
TPS–7 x TPS–67	244.7	0.048	1.49	58.3
MF– I x TPS–13	323.0	0.048	2.372	95.60
MF–II x TPS–13	318.5	0.048	1.242	49.82
TPS–7 x TPS–13	302.7	0.049	1.224	52.1
CD (0.05)	32.8	0.005	0.11	2.17

Significantly higher TPS yield was recorded in MF–I x TPS–13 (95.6 Kg/ ha), than other TPS families. The superiority of MF–I x TPS–13 might be due to higher number of berries and TPS production per plant. Suitability of combination of MF-I as female and TPS-13 as male parent was also reported by Vaghela *et al.,* (2007).

It can be concluded from the trial that after introduction and acclimatization for one season in the climate, the cross combination of MF-I x TPS-13 has shown better results in term of profuse flowering, berry setting, number of seeds per berry and TPS yield. Therefore, it may be recommended for commercial hybrid TPS production in the region.

4.2
Seed Germination and Vigour

(1) Potato fruits/berries

(2) True Potato Seed (TPS)

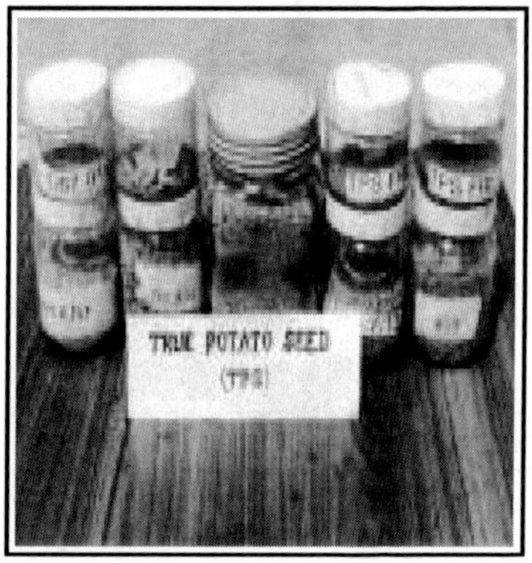

(3) Harvested berries

(4) TPS storage for trial

(5) Nursery raising in trench from TPS

Plate 6

Quality of True Potato Seed as Affected by Different TPS Families

Genetic constitution of male and female parental lines determines the TPS quality and quantity. Upadhya *et al.,* (1984) suggested that the selection of parental lines should be made on the basis of their yield potential, resistance to major diseases, ability to bloom under the conditions and set barries containing a high number of bold seeds. The erratic germination and low seedling vigour result in high mortality of the seedlings after transplantation. The good progeny should have attributes of high germination percentage, seedling vigour, withstanding transplanting shock, high yield potential and homogeneity for plant and tuber characteristics. Upadhya, (1983) reported that TPS with A type embryo give the highest and synchronous germination, vigorous seedlings and these seedlings give higher yield as compared to other embryo type seeds.

The study was conducted to evaluate TPS quality of six hybrid families i.e. MF–I x TPS–67, MF–II x TPS–67, TPS–7 x TPS–67, MF–I x TPS–13, MF–II x TPS-13 and TPS-7 x TPS–13. For evaluation of performance for percentage of embryo type (based on embryo-anatomical study as stated by Upadhya *et al.,* 1981), shoot length, root length and seedling vigour in the laboratory, the trial was conducted in randomized block design with three replications. Randomly selected 250 seeds of each TPS family were used per replication for the study. The hybrid TPS of the families were produced during the previous year at the research farm of the Institute. TPS of family MF–I x TPS–13 was also procured from CPRI-Campus Modipuram, Meerut (UP) and included in the study for comparative performance. True potato seeds having different type of embryos were categorized based on boldness *i.e.* A, B, C, D, E (A is having the highest boldness and D the least) identified with the help of stereoscopic binocular microscope (Fig. 4.2.1). Unidentified embryos of the seeds were kept under E category. Germination test was conducted in Petri–dish with moist germination paper. The paper was kept moist by regular watering and seed germination was recorded at 30 days after sowing. The seedling vigour index was computed by multiplication of germination percentage and total seedling length. Uniformity percentage of the seedlings was recorded on the basis of total uniform seedlings appearance of each family.

The results (Table 4.2.1) revealed that type A (circinate shaped) embryos were maximum in number in all the families. Among the families, MF–I x TPS–13 followed by MF–II x TPS–67 produced significantly higher number of type A embryos. Type B (Inverted U shaped) was the next most frequent class followed by D type (rod shaped) embryos. Type C (curved shaped) embryos were minimum in number in all the families. Unidentified or aborted embryos (type E) were present in all the families but frequency of this type embryos was lower with MF–I x TPS–13 followed by TPS–II x TPS–67 families. MF–I x TPS–67 was having maximum 14.4% E type embryos among the families. Over all Type A and Type B embryos together constituted maximum percentage of the families. Upadhya (1983) had described five types of embryos in TPS and suggested that embryo type can be used as a selection for suitable family, which could produce maximum bold TPS with vigorous growth of the seedling and higher survival percentage of the seedlings after transplantation. The result are in the agreement with the findings of Gopal (2003) who also reported 92-94% embryos together under Type A and Type B in various cross and open-pollinated TPS families.

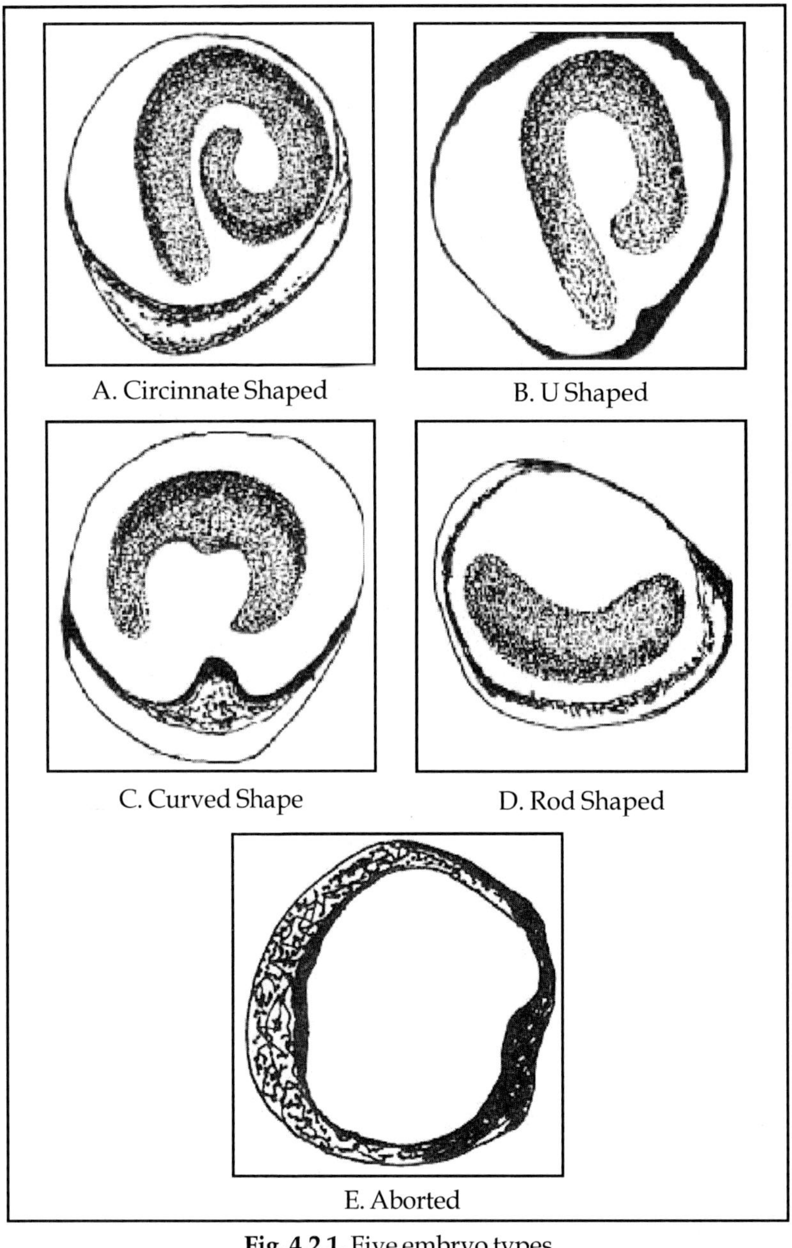

Fig. 4.2.1. Five embryo types
(*Source:* Advances in Horticulture Vol. 7).

Table 4.2.1: Percentage of TPS having different types of embryos of various hybrid TPS families.

Treatments	Type of embryo %				
	A	B	C	D	E
TPS Families					
MF – I x TPS – 67	51.2	14.2	4.6	12.6	14.4
MF – II x TPS – 67	72.4	10.0	4.0	7.4	6.2
MF – 7 x TPS – 67	44.0	17.2	4.4	15.8	13.6
MF – I x TPS – 13	78.8	12.0	3.6	7.6	3.6
MF – II x TPS – 13	55.6	14.0	6.4	12.4	11.6
MF – 7 x TPS – 13	49.2	20.2	5.0	14.0	7.6
MF – I x TPS – 13 M*	56.4	11.2	5.6	8.5	5.6

* The TPS was procured from CPRI, Modipuram

Significant difference among the TPS families was recorded in respect of germination percentage (Table 4.2.2). The family MF–II x TPS–67 was recorded higher (82.0%) germination which observed significantly superior on the other families. Maximum shoot and root length, and total length of the seedling were recorded with MF–I x TPS–13 family. Higher seedling vigour index was recorded with MF–I x TPS–13 family might be due to the family have maximum shoot and root length followed by MF–II x TPS–67. Maximum seedling uniformity percentage was observed with MF–II x TPS–67 followed by TPS–I x TPS–13 and both the family were observed statistically at par in respect of seedling vigour index and significantly superior than the other families. Thompson *et al.,* (1983) reported similar results on seed germination of the TPS families. Fast development of the root system had been associated with field establishment of the seedlings in other crops, too (Lee and Hackett, 1976; Esu, 1977; Yamaguchi and Street, 1977; Arteca, 1982; Struve and Moser, 1984).

Table 4.2.2: Germination & seedling vigour performance of different TPS families

Treatments	Germination (%)	Seedling Shoot length (cm)	Root length (cm)	Total length (cm)	Vigour index	Uniformity (%)
TPS families						
MF – I x TPS – 67	62.6	4.4	2.7	7.1	515.5	79.5
MF – II x TPS – 67	82.0	5.2	3.0	8.2	754.4	93.6
MF – 7 x TPS – 67	58.6	4.5	2.4	6.9	473.3	82.5
MF – I x TPS – 13	76.4	5.6	3.2	8.8	760.3	86.4
MF – II x TPS – 13	62.3	4.7	2.5	7.2	520.6	75.6
MF – 7 x TPS – 13	68.3	4.6	2.4	7.0	548.1	68.6
MF – I x TPS – 13 M*	78.2	4.0	2.1	6.1	477.0	77.9
CD (0.05)	2.8	0.42	0.35	0.65		

* The TPS was procured from CPRI, Modipuram

Gopal (2003) reported that in potato seedling vigour was a better indicator than seedling length for selection of suitable family for better survivability of the seedling. He observed seedling vigour alone could account for 45% of total variation for transplant survival, where as in combination with other character upto 48%. However, the other workers (Lee and Hackett, 1976; Esu, 1977; Yamaguchi and Street, 1977; Arteca, 1982; Struve and Moser, 1984) reported that all these characters were positively associated with each other for field establishment of the seedlings in other crops and can be considered for selection of suitable TPS families.

Maximum test weight (1000 seed wt.) of TPS was recorded with Type A followed by Type B embryos in all the families. Among the families, significantly higher average test weight was recorded with MF–I x TPS–13 (548 mg) followed by MF–II x TPS–67 (512 mg). TPS of other families produced in cold desert conditions (Leh) recorded higher test weight than the seed produced at in plains conditions (CPRI Campus Modipuram).

Table 4.2.3: 1000 TPS weight (mg) of different type of embryo

Treatments	Type of embryo					
	A	B	C	D	E	Average
TPS families						
MF – I x TPS – 67	658	522	304	443	402	466
MF – II x TPS – 67	726	619	333	445	388	512
MF – 7 x TPS – 67	636	558	318	456	356	465
MF – I x TPS – 13	817	736	312	488	388	548
MF – II x TPS – 13	615	553	323	410	376	455
MF – 7 x TPS – 13	650	535	326	436	362	462
MF – I x TPS – 13 M*	618	427	332	388	352	423

* The TPS was procured from CPRI Campus Modipuram, Meerut (U.P.)

The comparative data (Table 4.2.3 & Fig. 4.2.2) revealed that higher test weight of each type of embryos was recorded in MF–I x TPS–13 family when it was grown in natural climatic conditions of cold desert than grown in plains with artificially modified climatic conditions. Thus the study revealed that TPS family MF–I x TPS–13 followed by TPS–II x TPS–67 is most suitable to produce high quality TPS and potato crop under true potato seed technology programme and better quality TPS can be produced in cold desert than plain conditions.

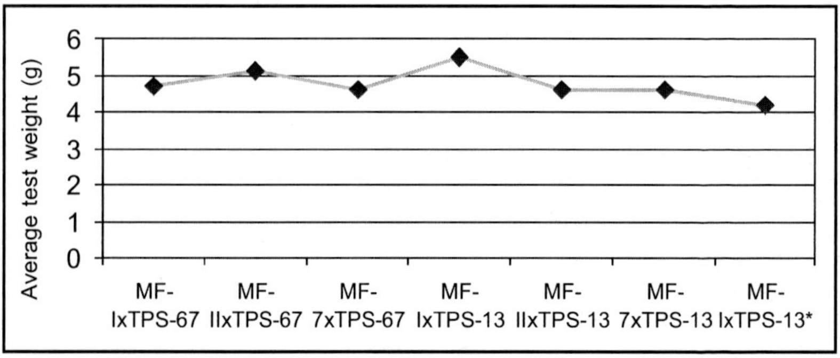

Fig.4.2.2. Average test weight of different TPS families produced in cold desert-Leh

□□□

4.3
Ware Crop through Seedlings

Transplantable size of TPS seedling in nursery beds

Transplantation of TPS seedlings in rows

Potato crop through TPS seedling transplantation

Harvested potato crop grown through TPS seedlings

Plate 7

Evaluation of Different TPS Families for Potato Tubers Production through Seedling Transplantation

The use of TPS as planting material has high production potential and there is no loss of potential food as planting material and during storage (Upadhya *et al.* 1986). Only 100-150 g TPS is sufficient to produce seedlings to transplant one hectare (Pandey *et al.* 2003) and that could reduce the cost of production considerably (Singh and Janardan Jee, 1990). At the above seed rate only 25-30 kg hybrid TPS is required as compared to 625-750 t seed tubers for planting 250 hectares of potato, which can help to make the Ladakh region self-sufficient in seed requirement.

The experiment was carried out during the cropping season (May–September) in 2002 and 2003. The TPS was sown in nursery beds in greenhouse in second week of March. TPS produced in the preceding year was used. To produce seedlings, the cultural practices as described in Extensions Bulletin No. 30 E of CPRI, Shimla were followed with slight modifications. The seedlings of six TPS families viz. MF-I/TPS-13 (I/13), MF-II/ TPS-13(II/13), MF-I/ TPS-67(I/67), MF-II/ TPS-67(II/67), TPS-7/ TPS-67(7/67) and MST-1 (OP) were transplanted in randomized block design with four replications in second week of May. The seedlings were transplanted with inter and intra row spacing of 50 cm x 20 cm, respectively in 4 m x 3 m size plots. The data were recorded on transplant survival (%) and plant height at haulm cutting, and total number of tubers and weight per plot in >40 g (large), 20-40 g (medium), and <20 g (small) grades and average tuber weight at harvest.

The highest survival percent of transplanted seedlings was obtained in TPS family I/13 (Table 4.3.1). The families II/67 and I/13 were at par for survival percent of seedling transplants and both were significantly superior than the other TPS families. Maximum plant height was recorded with OP seed of MST-1 but non-significant differences were recorded for plant height among all the families.

It is depicted from Table 4.3.2 that maximum total number of tubers (per plot) was recorded with I/13 and it was observed significantly higher over the other treatments. The lowest total number of tubers (per plot) was recorded with open pollinated TPS family MST-1. The TPS families II/13, I/67, II/67 and 7/67 have non-significant differences in respect of total number of tubers (per plot). The highest number of tubers in large (>40 g) and medium size (20-40 g) tuber grades was recorded from I/13 followed by 7/67. Among all the families, MST-1 (OP) recorded the lowest number of large and medium size tuber grades. TPS family I/13 followed by II/67 recorded maximum number of tubers under small size (<20 g) tuber grade but non-significant difference was observed between the families.

Table 4.3.1: Performance of different TPS families for percent survival, plant height, average tuber weight and tuber yield (pooled over 2002-03).

Treatments	Plant survival (%)	Plant height (cm)	Av. tuber wt. (g)	Tuber yield (q/ha)
TPS families				
I / 13	86.8	76.8	34.1	226
II / 13	73.8	70.5	27.9	148
I / 67	77.1	72.8	28.5	160
II / 67	82.2	77.8	23.5	137
7 / 67	75.1	75.4	35.6	197
MST – 1	74.5	79.2	21.8	113
C. D. (0.05)	5.8	9.3	5.8	21.5

Table 4.3.2: Total number of tubers and number of tubers in different grades of various TPS families (pooled over 2002-03)

Treatments	Total no. of tubers (per plot)	No of tubers/plot (> 40 g)	No of tubers/plot (20 – 40 g)	No of tubers/plot (< 20 g)
TPS families				
I / 13	898	226	280	406
II / 13	721	154	196	358
I / 67	755	140	235	376
II / 67	773	165	194	398
7 / 67	750	212	252	342
MST – 1	621	123	150	349
C. D. (0.05)	80.3	40.1	55.7	76.2

Results presented in Table 4.3.3 indicated that the TPS families have significant influence on total weight of tubers and weight of tubers under different grades. Maximum total weight of tubers was recorded in I/13 (30.1 kg/plot) followed by 7/67 (26.2 kg/plot). The total weight recorded in I/13 was significantly higher than all the other families. Higher tuber weight under large size grade (> 40 g) was recorded in case of 7/67 (17.7 kg/plot). However, 7/67 and I/13 were found at par and both were significantly superior over II/13, I/67, II/67 and MST-1. Higher weight of tubers under medium size (20-40 g) tuber grade was recorded in I/13 (9.0 kg/plot) and it has been shown significant superiority over all the other families. Although, maximum weight of tubers recorded under small size (<20 g) grade in I/13 (4.9 kg/plot) but non significant influence observed among the families. The OP family MST-1 was recorded the lowest total tuber weight and weight of tubers under all the three grades.

Table 4.3.3. Total weight of tubers and weight of tubers in different grades of various TPS families

Treatments	Wt. of total tubers (kg/plot)	Wt. of tubers kg/plot (> 40g)	Wt. of tubers kg/plot (20 – 40 g)	Wt. of tubers kg/plot (< 20 g)
TPS families				
I / 13	30.1	16.1	9.0	4.9
II / 13	19.6	9.6	5.8	4.4
I / 67	21.45	11.2	5.8	3.6
II / 67	16.85	8.6	5.3	4.4
7 / 67	26.2	17.7	5.9	3.5
MST – 1	14.7	7.4	4.8	2.4
C. D. (0.05)	3.7	2.08	1.5	2.3

The different TPS families had significant effect on average tuber weight and tuber yield (Table 4.3.1). The highest tuber weight was recorded in 7/67 which was at par with I/13 and significantly higher than II/13, I/67, II/67 and MST–1 families. Maximum tuber yield was recorded 226 q/ha in I/13 followed by 197 q/ha in 7/67 (Fig. 4.3.1). The tuber yield recorded with I/13 was significantly higher than all other families.

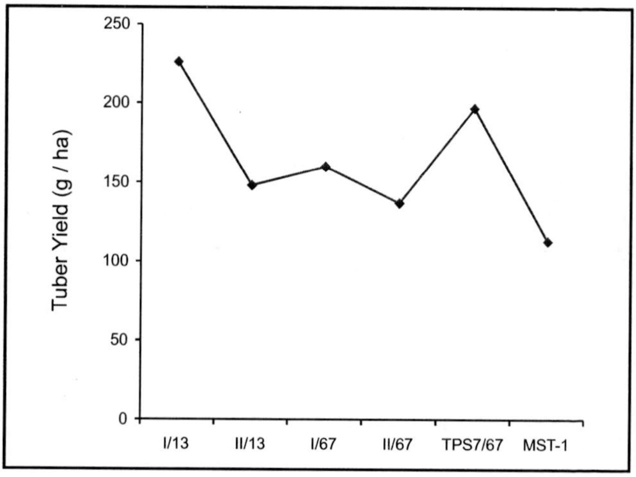

Fig.4.3.1. Average tuber yield of different TPS families

The results discussed above indicated that the TPS family I/13 is superior to other families for number, tuber weight and tuber yield per unit area and % marketable tubers therefore, it is more suitable to grow ware potato crop through seedling transplantation in the region.

Performance of TPS Families Under Different Spacing through Transplantation

Among the several factors of production, plant population per unit area and the TPS families play vital role in determining the quantity, quality and economics of the produce. High plant density adversely affects the quality of produce and too low plant density may also not be remunerative. The present investigation was planned to study the performance of various hybrid TPS families with different planting spacing.

The experiment was conducted from May–September in 1996 and 1997 at the research farm of DIHAR. Eighteen treatment combinations consisting six hybrid TPS families i.e. I/13 (V_1), II/13 (V_2), I/67 (V_3), II/67 (V_4), 7/13 (V_5) and 7/67 (V_6) and three spacing 45x10 cm (S_1), 45x15 cm (S_2), 45x20 cm (S_3) were distributed in randomized block design with three replications. Seedlings were raised in greenhouse using TPS of last year. TPS was sown in second week of March in nursery bed and seedlings transplanted at 10-12 cm height (60 DAS). The seedlings were transplanted in furrow. The furrow was irrigated two days before transplanting. The furrow was made in the plots measuring 1.20x1.20 m. Irrigation was applied immediately after transplanting and subsequently as when required. 50 kg, 80 and 60 kg / ha N, P_20_5 and K_2O was supplied at the time of last ploughing of the field, respectively and 50 kg N was given in two equal split doses at the time of earthing up. First earthing up was done at 35 days after transplanting and second at 60 days after transplanting. The haulms were removed 120 DAP during second week of September and crop was harvested after 12 days of haulm cutting. The produce of each plot was graded into three sizes *i.e.* >40 g (large), 20-40 g (medium) and <20 g (small). Average tuber weight was calculated. Total tuber yield q/ha was calculated on the basis of 10% area leaving under layout of the field. >20 g size tubers were considered for marketable yield. The two years data was pooled and statistically analyzed for treatment comparison.

The results from Table 4.3.4 show that the various TPS families were not influenced significantly in respect of total number of tubers per plot and number of tubers in difference grades (large, medium and small). However, total number of tubers per plot and under different grades were significantly influenced with spacing treatments. Significantly more number of total tubers (per plot) were recorded with close (45x10 cm) spacing and the lowest number of tubers were recorded with wider spacing (45x20 cm). More number of tubers in large (> 40g) and medium (20-40g) grades tubers were also recorded with wider spacing (45x20 cm). However, close spacing (45 x 10 cm) produced maximum number of small (<20 g) size tubers, than wider (45 x 15 cm and 45 x 20 cm) spacings.

The data presented in Table 4.3.5 revealed that TPS family II/67 produced the highest total tuber weight (kg/plot) but it was not significantly different from other families except II/13 which produced lowest total tuber weight (kg/plot). Spacing treatments significantly influenced total tuber weight (kg/plot). Maximum total tuber weight (kg/plot) was recorded with 45x20 cm spacing which was significantly higher than other spacing treatments. The spacing 45x10 and 45x15 cm remained at par for total tuber weight (kg/plot). Significantly more weight under large size (>40g) grade tubers was recorded with TPS family II/67. The wider spacing (45x20 cm) yielded more large size tubers (>40g) than other spacing treatments. The highest medium size (20-40g) tuber weight was recorded in

II/67 followed by 7/67 and both were superior to other families but were at par with each other. For medium size tubers (20-40g), higher tuber weight was recorded with 45x20 cm spacing and it was significantly higher than other spacing treatments. Higher tuber weight in small size tuber (<20g) grade was recorded in I/67 than II/13, II/67 and 7/67. TPS families I/13 and 7/67 recorded similar weight of tubers in small size (<20 g) grade and both were at par with I/67. Close spacing yielded significantly more tubers under small size grade.

Table 4.3.4: Total number of tubers and number of tubers in different grade as influenced by hybrid TPS families and spacing (pool over 1996-97)

Treatment	Total no. of tubers /plot	No of tubers /plot (>40g)	No. of tuber /plot (20 – 40g)	No. of tuber /plot (<20g)
TPS family				
I / 13 (V_1)	137.6	37.1	48.4	51.1
II / 13 (V_2)	137.9	39.3	50.9	48.1
I / 67 (V_3)	155.9	45.6	52.8	61.3
II / 67 (V_4)	138.8	39.5	50.7	52.8
7 / 13 (V_5)	165.2	42.7	54.0	63.4
7 / 67 (V_6)	160.1	44.8	62.8	71.1
CD (0.05)	31.1	9.6	17.1	21.1
Spacing (cm)				
45 x 10 (S_1)	153.2	39.2	54.5	68.6
45 x 15 (S_2)	152.8	40.5	54.7	59.7
45 x 20 (S_3)	139.2	47.5	65.7	50.6
CD (0.05)	14.4	7.8	12.1	8.9

Table 4.3.5: Total number of tubers and number of tubers in different grade as influenced by hybrid TPS families and spacing (pooled over 1996-97)

Treatment	Total Wt. of tubers (kg/plot)	Wt. of tubers kg/plot (>40g)	Wt. of tubers kg/plot (20 – 40g)	Wt. of tubers kg/plot (<20g)
TPS family				
I / 13 (V_1)	4.1	2.6	1.21	0.73
II / 13 (V_2)	3.7	2.6	1.25	0.70
I / 67 (V_3)	3.9	2.5	1.34	0.82
II / 67 (V_4)	4.5	3.3	1.58	0.69
7 / 13 (V_5)	3.8	2.5	1.17	0.65
7 / 67 (V_6)	4.2	2.5	1.56	0.73
CD (0.05)	0.5	0.3	0.16	0.12
Spacing (cm)				
45 x 10 (S_1)	3.4	2.2	1.16	0.83
45 x 15 (S_2)	3.9	2.6	1.27	0.70
45 x 20 (S_3)	4.8	3.0	1.57	0.58
CD (0.05)	0.4	0.3	0.18	0.12

Table 4.3.6 showed that average tuber weight, tuber yield and marketable tuber yield were influenced significantly by the TPS families and spacing treatments. Maximum 40.6 g average tuber weight was recorded in II/67, which was significantly higher than other TPS families. Wider spacing (45x20 cm) was significantly superior among other spacing treatments and recorded maximum 32.2 g average tuber weight under the treatment. Higher average tuber weight of potato in wider spacing has also been reported by Singh *et al.,* (1997). Maximum tuber yield (331 q/ha) was recoded in II/67 and it was statistically at par with I/13 and 7/67 but significantly superior to II/13, I/67 and 7/13. Higher tuber yield (350 q/ ha) was recorded with 45x20 cm spacing and was significantly

superior than 45x15 cm and 45x10 cm spacing treatments. Higher marketable tuber yield was recorded in II/67, which was significantly higher than I/13, I/67, 7/13 and 7/67 but was statistically at par with II/13. Significant higher marketable tuber yield was recorded under 45x20 cm than 45x15 cm spacing treatments.

Table 4.3.6. Effect of different hybrid TPS families and spacing on average tuber weight, tuber yield and marketable tuber yield (pooled over 1996-97).

Treatment	Av. tuber wt. (g)	Tuber yield (q/ha)	Marketable tuber yield (%)
TPS family			
I / 13 (V_1)	36.3	315	84.9
II / 13 (V_2)	31.9	279	87.3
I / 67 (V_3)	34.2	284	82.6
II / 67 (V_4)	40.6	331	90.1
7 / 13 (V_5)	30.5	267	83.3
7 / 67 (V_6)	33.0	304	81.6
CD (0.05)	2.3	30.2	3.0
Spacing (cm)			
45 x 10 (S_1)	20.4	244	79.5
45 x 15 (S_2)	27.7	295	85.3
45 x 20 (S_3)	32.2	350	88.9
CD (0.05)	2.2	21.4	2.1

Interaction of TPS families and planting spacing was observed to vary according to the TPS families. Total number of tubers and tuber number in different grades (per plot) was influenced by the TPS families and spacing treatments (Table 4.3.7). Maximum total number of tubers was recorded in 7/13 planted at 45x10 cm spacing followed by 7/67 at 45x15 cm and was significantly higher than all the treatment combinations except V_6S_3, V_3S_2 and V_1S_2. The more number of large size (>40g) tubers was recorded in I/67 and 45x15 cm

(V_3S_2) followed by II/67 and 45x20 cm (V_4S_3) treatment combination but was non significantly higher than other treatment combinations. Similarly the highest number medium and small size tubers were produced in 7/67 at 45x20 cm (V_6S_3) and 7/13 at 45x10 cm (V_5S_1) treatment combinations, respectively and both the treatments recorded non-significant differences over the other treatment combinations.

Table 4.3.7: Interaction of hybrid TPS families and spacing on total number and number of tubers in different grades (per plot)

Spacing (cm)	TPS Families						
	I/13 (V_1)	II/13 (V_2)	I/67 (V_3)	II/67 (V_4)	7/13 (V_5)	7/67 (V_6)	Mean
	(1) Total no. of tuber/plot						
45 x 10 (S_1)	138.6	149.8	148.1	143.0	198.8	141.3	153.3
45 x 15 (S_2)	166.6	122.3	178.3	128.9	139.6	180.9	152.8
45 x 20 (S_3)	107.8	141.6	141.3	144.5	142.3	158.1	139.3
Mean	137.7	137.9	155.9	138.8	160.2	160.1	
	(2) No. of tubers/plot (>40g)						
45 x 10 (S_1)	38.3	41.3	39.1	35.1	45.8	36.4	39.3
45 x 15 (S_2)	40.9	36.6	50.3	35.4	40.1	39.5	40.5
45 x 20 (S_3)	32.1	39.9	47.5	48.1	43.3	43.6	42.4
Mean	37.1	39.3	45.6	39.5	43.1	39.8	
	(3) No. of tubers/plot (20-40g)						
45 x 10 (S_1)	46.6	57.9	50.9	54.0	60.6	56.9	55.0
45 x 15 (S_2)	61.5	42.6	60.0	46.6	49.1	63.6	53.9
45 x 20 (S_3)	37.1	52.1	47.5	51.6	52.3	68.3	51.5
Mean	48.4	50.9	52.8	50.7	54.0	63.0	
	(4) No. of tubers/plot (<20g)						
45 x 10 (S_1)	53.6	50.4	66.3	53.5	93.1	64.5	63.6
45 x 15 (S_2)	64.1	44.6	71.3	46.8	50.3	81.3	59.7
45 x 20 (S_3)	35.6	49.4	46.3	57.9	46.6	67.8	50.6
Mean	51.1	48.1	61.3	52.7	63.3	71.2	

CD (0.05)	(1)	(2)	(3)	(4)
S x F	36.6	19.3	29.6	36.5

The data in Table 4.3.8 revealed that combined effect of TPS families and spacing treatments has significant influence on total tuber weight and weight of tubers in different grades(kg/plot). Maximum total tuber weight was recorded in II/67 and 45x20 cm (V_4S_3) treatment combination but was recorded statistically at par with V_1S_3 and V_6S_3 and significantly superior over other treatment combinations. The higher tuber weight of large and medium size grades was also recorded with II/67 and 45x20 cm (V_4S_3) treatment combination. However, maximum weight of tubers under small size tuber grade was recorded with II/13 and 45x10 cm (V_2S_1) treatment combination and it was statistically at par with all the families planted at 45x10 cm spacing but significantly superior than all the other treatments.

Table 4.3.8: Interaction of hybrid TPS families on total tuber weight (per plot) and weight of tubers in different grade (pooled over 1996-97)

Spacing (cm)	TPS Families						
	I/13 (V_1)	II/13 (V_2)	I/67 (V_3)	II/67 (V_4)	7/13 (V_5)	7/67 (V_6)	Mean
	(1) Total wt. of tuber/plot						
45 x 10 (S_1)	3.3	3.4	3.3	3.7	3.5	3.3	3.4
45 x 15 (S_2)	4.0	3.5	3.9	3.7	3.8	4.4	3.9
45 x 20 (S_3)	5.0	4.3	4.5	6.0	4.1	4.7	4.8
Mean	4.1	3.7	3.9	4.5	3.8	4.1	
	(2) Wt. of tubers/plot (>40g)						
45 x 10 (S_1)	2.1	2.4	2.1	2.2	2.3	1.6	1.8
45 x 15 (S_2)	2.0	2.5	2.8	2.9	2.5	2.3	2.5
45 x 20 (S_3)	3.3	2.8	2.7	3.5	2.8	3.1	3.0
Mean	2.5	2.6	2.5	2.9	2.5	2.3	
	(3) Wt. of tubers/plot (20-40g)						
45 x 10 (S_1)	1.1	1.3	1.2	1.6	1.2	1.1	1.3
45 x 15 (S_2)	1.2	1.0	1.3	1.4	1.2	1.6	1.3
45 x 20 (S_3)	1.4	1.5	1.6	2.1	1.2	1.7	1.6
Mean	1.2	1.3	1.4	1.7	1.2	1.5	

contd...

table contd...

			(4) Wt. of tubers/plot (<20g)				
45 x 10 (S_1)	0.9	0.9	0.8	0.9	0.8	0.8	0.9
45 x 15 (S_2)	0.8	0.6	0.8	0.7	0.7	0.7	0.7
45 x 20 (S_3)	0.6	0.6	0.6	0.6	0.5	0.6	0.6
Mean	0.8	0.7	0.7	0.8	0.7	0.7	

CD (0.05)	(1)	(2)	(3)	(4)
S x F	0.9	0.7	0.4	0.2

Interaction of TPS families and spacing treatments on average tuber weight, tuber yield and marketable tuber yield was observed (Table 4.3.9). Maximum average per tuber weight was recorded with II/67 and 45x20 cm (V_4S_3) and was at par with I/13 and 45x20 cm (V_1S_3) and significantly superior over other treatments. Higher tubers weight under the treatments may be due to the reason that the treatments have maximum large size tubers. The highest tuber yield 421 q/ha was recorded of II/67 and 45x20 cm (V_4S_3) followed by 345 q/ha in 7/67 and 45x20 cm (V_6S_3) and was significantly higher than all other treatments.

Table 4.3.9: Interaction of hybrids TPS families and spacing on average tuber wt., tuber yield and marketable tuber yield (pooled over 1996-97)

Spacing (cm)	**TPS Families**						
	I/13 (V_1)	**II/13 (V_2)**	**I/67 (V_3)**	**II/67 (V_4)**	**7/13 (V_5)**	**7/67 (V_6)**	**Mean**
			(1) Av. per tuber wt. (g)				
45 x 10 (S_1)	23.9	41.7	29.3	29.7	27.0	29.4	30.3
45 x 15 (S_2)	32.7	32.5	32.0	35.5	28.9	32.5	32.3
45 x 20 (S_3)	47.2	35.7	41.4	47.6	35.6	34.7	40.4
Mean	34.6	37.0	34.2	37.6	30.5	32.2	
			(2) Tuber yield (q/ha)				
45 x 10 (S_1)	268.4	237	236	243	229	252	244
45 x 15 (S_2)	297.6	265	294	328	274	314	295
45 x 20 (S_3)	279.6	334	322	421	299	345	333
Mean	282.0	279	284	331	267	304	

contd...

table contd...

	(3) Marketable tuber yield (%)						
45 x 10 (S_1)	80.0	84.1	77.8	82.0	78.3	74.5	79.5
45 x 15 (S_2)	85.7	87.6	83.5	89.8	84.2	81.4	85.4
45 x 20 (S_3)	89.1	90.2	86.5	92.2	87.4	88.9	89
Mean	84.9	87.3	82.6	87.7	83.3	81.6	
CD (0.05)	(1)			(2)		(3)	
S x F	5.6			52.4		2.7	

The higher percentage of marketable tuber was also recorded in II/67 family and 45x20 cm (V_4S_3) but was at par with II/13 and 45x20 cm (V_2S_3) and significantly more than all other treatments. The treatment V_4S_3 has more tubers under large and medium size grades, may be due to this reason the treatment has maximum marketable tuber yield.

In this trial more number of large size tubers were obtained with wider spacing therefore, higher tuber yield of all the families was obtained with 45x20 cm spacing and with this spacing TPS family II/67 produced higher tuber yield. Therefore, it can be concluded that maximum tuber yield with higher average tuber weight can be produced by planting of seedlings of II/67 family at 45x20 cm spacing.

□□□

4.4
Production of Seedling Tubers

Plate 8: Different grades of seedling tubers

Evaluation of Different TPS Families for Seedling Tubers Production in Nursery Bed

The production of seedling tubers from TPS in first year and planting them to grow ware potato crop in following season appears promising. Hence, the trials were conducted to ascertain the performance of various hybrid TPS progenies for seedling tuber production.

The trials were carried out during the cropping season (May–September) of 1997 and 1998 in randomised block design with three replications having a net plot size of 1 m x 1 m. TPS of seven progenies was produced in 1996 at the Research Farm of DIHAR. Seeds of seven TPS genotypes i.e. HPS-I/13, HPS–II/13, HPS–I/67, HPS- II/67, HPS 7/67, II/TPS–FRL–1 and TPS–FRL–2/67 were sown in nursery beds in an under ground greenhouse in first week of April during both the years 3 kg/m^2 well-decomposed FYM was applied 15 days before the transplantation of the seedlings. Seedlings at 8–10 cm, height were transplanted in open field nursery beds in the first week of June at 25x4 cm spacing (100 plant/m^2). The crop was harvested in third week of September during both the years. The data on survival of plants (%), number of tubers/m^2, total tuber yield (kg/m^2) and average tuber weight (g) were recorded at the time of harvest. The produce of each plot was graded into three sizes i.e. large (>35 g), medium (20–35 g) and small (<20 g). Number and weight of each grade was recorded separately. The data of both the years was pooled and analyzed.

Percentage survival of the seedlings in the nursery bed ranged from 87.0 to 97.4. TPS–FRL–2/67 progeny had the highest survival which was at par with HPS II/67, II/TPS-FRL-1 and HPS II/13 and significantly higher than other progies (Table 4.4.1). From the results it can be concluded that good survival of progeny produce more number of tubers and tuber yield / m^2 Pandey *et al.* (1990). Thre better adaptability and high seedlings survival rate of TPS-FRL-2/67 could be attributed to the TPS-FRL-2, one of the parent of the progeny which was developed through clonal selection of local germplasm.

Table 4.4.1: Performance of different TPS progenies for seedling survival, number of tubers/m^2 and tuber yield (kg/m^2) (pooled over 1998-99).

TPS Progenies	Seedling survival %	No. of tubers/m^2	Tuber yield kg/m^2
HPS – I/13	87.4	213	4.3
HPS – II/13	91.3	261	5.4
HPS – I/67	87.7	244	5.1
HPS – II/67	95.4	287	6.6
HPS – 7/67	86.9	238	4.2
II/TPS–FRL –I	95.3	259	5.6
TPS – FRL – 2/67	97.4	325	7.2
CD (0.05)	8.0	68.3	1.3

Data presented in Table 4.4.1 revealed that the total number of tubers produced by the progenies ranged from 213 to 325 per m2. TPS-FRL-2/67 produced highest number of seedling tubers followed by HPS-II/67 and was significantly higher than HPS-I/13, HPS-7/67 and HPS I/67 and at par with other progenies. Among the hybrid TPS progenies HPS-7/67 recorded the lowest seedling survival rate which might be reason for the lowest number of tubers.

TPS-FRL-2/67 produced maximum tuber yield (7.2 kg/m^2) followed by TPS-II/67 (6.6kg/m^2). It is depicted from Table 4.4.1 that both the families had showed better seedling survival. Minimum average tuber yield (4.2 kg/m2) was observed in HPS–7/67, which could be due to the lowest seedling survival in this family.

Table 4.4.2: Average tuber weight, weight and number of tubers (> 35 g) in different TPS families (pooled over 1998-99).

TPS Progenies	Av. tuber wt. (g)	Tuber wt. kg/ m² (>35g)	No. of tuber/ m² (>35g)
HPS - I/13	21.8	1.7	27.3
HPS - II/13	21.1	2.1	29.8
HPS - I/67	21.4	1.3	29.8
HPS - II/67	23.1	2.5	46.0
HPS - 7/67	18.0	1.8	29.1
II/TPS–FRL –I	21.8	2.1	49.5
TPS - FRL - 2/67	22.2	2.6	59.6
CD (0.05)	6.0	0.7	5.4

The data presented in Table 4.4.2 and Table 4.4.3 indicate that the highest number of tubers were produced from small size (< 20 g) seedling tuber grade in all the progenies. The maximum number and weight per m² of large (>35g) and medium (20-35g) size tubers was recorded in TPS-FRL-2/67.

Table 4.4.3: Total tuber weight and numbers of different grade of TPS families (pooled over 1998-99)

TPS Progenies	Tuber wt. kg/m² (20 – 35g)	No. of tuber/m² (20 – 35g)	Tuber wt. (<20 g)	No. of tuber/m² (<20 g)
HPS - I/13	1.2	40.6	1.3	144.8
HPS - II/13	1.5	49.6	1.7	181.3
HPS - I/67	1.7	56.8	1.5	156.9
HPS - II/67	2.0	67.3	2.1	173.8
HPS - 7/67	1.3	42.1	1.2	167.0
II/TPS–FRL –I	2.1	85.1	1.1	123.0
TPS - FRL - 2/67	3.2	117.3	1.3	151.3
CD (0.05)	0.5	13.4	0.6	65.7

TPS FRL-2/67 and II / TPS-FRL-1 had higher number and more weight of medium (20-35g) size tubers than all other TPS families. It indicates that these families have maximum number of tubers in large and medium size grade in comparison to other families. The difference between number of tubers in between different grades was much higher in all the families but by weight it was at par.

On the basis of above findings it may be concluded that the family TPS- FRL-2/67 followed by HPS II/67 can be used as seedling tuber production and these seedling tubers may be used as quality seed for the production of ware potato crop in cold desert agro-climatic conditions of the region.

❑❑❑

4.5
Crop through Seedling Tubers

Potato crop through seedling tubers

Harvested potato crop grown through seedling tubers

Plate 9

Performance of Different TPS Families for Potato Production through Seedling Tubers

Seedling tubers are produced either from direct seeding of TPS in the nursery beds or by transplanting seedlings from nursery to the field. At maturity, tubers are harvested, graded and stored till next planting season. This method combines the attributes of potato production from TPS and those of seed tubers. In this system seedling tubers can be produced in small protected areas where adverse conditions can properly be managed. However, ware potato crop through seedling transplanting can also be taken successfully in the region but in earlier trial its low yield has been observed in comparison to traditional crop and the crop produced from direct transplantation generally produce large number of small size tubers which may not be useable. In other hand seedling tubers produced in nursery bed can overcome the problem. Potato crop through seedling tubers also can be taken as crop from seed tubers as farmers are familiar with the method. Therefore, to find out suitable hybrid TPS strain for maximum yield, the present investigation was carried out with small size (20-25 g) tubers used as planting material.

The experiment was laid out in randomized complete block design and four replications with five hybrid TPS families *i.e.* I/13, II/13, I/67, II/67 and 7/67 along with traditionally grown local variety as control during the cropping season (May-September) in 2002 and 2003. The seedling tubers of 20-25 g size produced in previous year were planted in plot size of 2.0 m x 1.6 m with inter and intra row spacing of 50 cm and 20 cm, respectively. The seedling tubers were planted in second week of May during both the years. All other cultural practices were followed as practiced in the region. The crop was dehaulmed at 120 days after planting (second week of September) and harvested after 15 days. Data were recorded at the time of harvesting on total number of tubers and weight/ plot for different grades in >40 g (large), 20-40 g (medium) <20 g (small).

From the pooled data of two years (Table 4.5.1) it was clear that the total number of tubers (per plot) and number of tubers (per plot) in different grades were influenced by TPS families. TPS family II/67 followed by 7/67 produced significantly more number of tubers (per plot) than other TPS families and local check. The similar trend was observed for large tuber grade. From medium and small tubergrade higher number of tubers were produced in II/67, which was at par with 7/67 but significantly higher than the other families and local variety (control).

Table 4.5.1: Effect of different TPS families on total number of tubers and number of tubers/plot in different grades (pooled over 2002-03)

Treatment	Total no. of tubers/plot	No. of tubers/plot (>40g)	No. of tubers/plot (20-40g)	No. of tubers/plot (<20g)
I/13	258.5	91.0	81.7	83.1
II/13	210.0	101.2	63.6	82.3
I/67	298.1	110.7	87.5	93.2
II/67	349.0	127.2	115.1	102.1
7/67	324.2	112.3	107.8	98.4
Local	263.9	104.8	105.5	81.4
CD (0.05)	21.92	6.80	6.52	7.5

All the TPS families produced more total tuber weight (kg/plot) than local variety (Table 4.5.2). Significantly higher total tuber weight was recorded with II/67 (12.95 kg/plot) followed by I/13 (11.34 kg/plot). However, TPS families II/13, I/67 and 7/67 yielded at par with local variety for total weight of tubers (per plot). The similar trend was observed for >40 g (large) size tuber grade. The weight of tubers of II/67 and I/13 was almost identical and both the families were significantly superior than others and local variety in respect of medium size (20-40 g) tuber grade. The difference in weight of tubers was found non-significant among the families for small size (<20 g) tuber grade.

Table 4.5.2: Effect of different TPS families on total weight of tubers and weight of tubers/plot in different grades (pooled over 2002-03) at par with 7/67

Treatment	Total Wt. of tubers (kg/plot)	Wt. of tubers >40g (kg/plot)	Wt. of tubers 20-40g (kg/plot)	Wt. of tubers <20g (kg/plot)
I/13	11.34	7.7	1.90	1.03
II/13	7.98	5.4	1.32	1.06
I/67	8.21	5.8	1.49	0.93
II/67	12.95	9.5	1.91	1.22
7/67	8.22	5.5	1.81	1.06
Local	7.34	4.7	1.36	0.93
CD (0.05)	1.07	0.52	0.27	0.34

It is evident from the Table 4.5.3 that among the TPS families and local variety, significantly higher average tuber weight was recorded with I/13. All the TPS families produced more tuber yield (q/ha) than local variety. Maximum tuber yield (364 q/ha) was recorded in II/67 followed by I/13 (329 q/ha). It was found significantly higher than all the other TPS families.

Table 4.5.3: Effect of different TPS families on average tuber weight and yield (pooled over 2002-03)

TPS families	Av. tuber weight (g)	Tuber yield (q/ha)
I/13	45.8	329
II/13	38.1	225
I/67	28.1	231
II/67	37.8	364
7/67	25.5	235
Local	28.5	211
CD (0.05)	3.1	35.0

From the results it may be concluded that seed rate may be reduced by using small size seedling tubers of TPS family and maximum tuber yield than local variety can be obtained with seedling tubers of TPS family II/67 for ware potato crop production in high altitude cold desert climatic conditions. TPS technology has the potential to fulfill the local requirement of quality planting material and production of ware potato crop through TPS derived planting material in remote areas like Ladakh.

❑❑❑

4.6
Evaluation of Promoters for Yield

Performance of TPS Families with Different Spacings for Potato Production through Seedling Tubers

Earlier trials have shown that potato cultivation from TPS derived seedling tubers is viable. The mechanism of raising ware crop from seedling tubers is similar to that of traditional system using seed tubers. Seedling tubers produced under the proposed technology can fulfill the requirement of quality planting material of the region. Plant stand has a great influence on quantity, quality and economics of the produce. The present study was intended to determine the potential of seedling tubers of various hybrid TPS families and standardization of spacing for ware potato production in the long day conditions of the region.

The experiment was carried out during 1996 and 1997. The fifteen treatments comprising of five hybrid TPS families I/13 (V_1), II/13 (V_2), I/67 (V_3), II/67 (V_4) and 7/67 (V_5) and three spacing combinations 50x10 cm (S_1), 50x15 cm (S_2) and 50x20 cm (S_3) were laid out in factorial randomized block design with three replications. 30-40g seedling tubers of each TPS family were planted. The other practices were followed as described in earlier experiment. The crop was dehaulmed in second week of September (120 DAP) and harvested after another 15 days. The produce of the plots was graded as >40g, 20-40g and <20g tuber size. The number and weight of all the tuber grades were recorded and pooled data analyzed.

Table 4.6.1: Effect of TPS families and spacing on number of tubers under different grades (pooled over 1996-97)

Treatment	Total no. of tubers/plot	No of tubers/plot (>40g)	No. of tubers/ plot (20-40g)	No. of tubers/plot (<20g)
TPS families				
I/13	287	97.7	96.6	87.8
II/13	316	87.1	118.3	125.2

contd...

table contd...

I/67	317	96.7	104.3	103.1
II/67	347	112.5	118.0	132.9
7/67	353	92.8	115.1	146.5
CD (0.05)	49.7	14.0	26.9	35.0
Spacing				
50 x 10	320	91.0	110.4	116.4
50 x 15	338	108.1	117.3	127.2
50 x 20	310	90.0	101.8	113.7
CD (0.05)	24.5	15.0	11.8	27.1

Data indicated that maximum total number of tubers were observed with TPS family 7/67 (Table 4.6.1). Significantly the lowest total number of tubers were recorded in I/13 and the other TPS families were statistically at par. Significantly maximum number of tubers were recorded with II/67 in large size (>40 g) tuber grade whereas, TPS family I/13 recorded the lowest number of tubers in all the three grades. The other TPS families did not differ significantly each other for medium (20-40 g) and small (<20 g) size tubers. Higher total number of tuber were recorded at 50x15 cm spacing and they were statisticaly at par with the tubers at 50x10 cm spacing and significantly more than 50x20 cm spacing treatment. The simlar trend was observed for large (> 40 g) and medium (20-40 g) size tuber grade. However, difference in number of tubers under small (< 20 g) size tubes grade was non-significant at all the three spacing treatments.

Table 4.6.2: Performance of weight of tubers under different grades with various TPS families and spacing (pooled over 1996-97)

Treatment	Total wt. of tubers kg/plot	Wt. of tubers kg/plot (>40g)	Wt. of tubers kg/plot (20-40g)	Wt. of tubers kg/plot (<20g)
TPS families				
I/13	13.73	9.78	3.19	1.02
II/13	13.29	8.98	4.04	1.54
I/67	14.48	10.34	3.35	1.37

contd...

table contd...

II/67	15.72	11.11	4.56	1.72
7/67	14.27	9.39	3.82	1.60
CD (0.05)	1.5	1.20	0.97	0.50
Spacing				
50 x 10	13.59	9.17	3.79	2.46
50 x 15	15.84	11.48	4.07	2.58
50 x 20	13.46	9.10	3.53	1.31
CD (0.05)	1.1	1.50	0.75	0.40

Total tuber weight and weight of tubers (kg/plot) in different grades such as large, medium and small was influenced significantly by the different TPS families and spacing (Table 4.6.2). Maximum total weight of tubers was recorded for TPS family II/67 and was statisticaly at par with I/67 and 7/67 but significantly higher than all the other famlies. From the large size tuber grade, maximum weight of tubers was recorded for II/67, which was at par of I/67 but significantly higher than all the other families. Higher weight of tubers from medium size tuber grade was recorded for II/67 and it was at par of II/13 and 7/67 but statisticaly significantly than all the other TPS families. TPS family I/13 yielded significantly lower weight of tubers under small (< 20 g) size tuber grade and all the other families were at par in respect of weight of tubers under this grade. Significantly higher total weight of tubers was recorded at 50x15 cm followed by 50x10 cm spacing treatment. Similar trend was observed for large (> 40 g) size tuber grade. Non-significant difference for weight of tubers was recorded among all the spacing treatment for medium (20-40 g) size tuber grade. However, significantly lower weight of tubers was recorded at wider (50x20 cm) spacing treatment and other spacing treatments (50x10 cm, and 15x15 cm) were at par in respect of weight of tubers under this grade.

Different TPS families and spacing treatments had significant effect on average tuber weight, tuber yield and marketable tuber yield (Table 4.6.3). The maximum average tuber weight was recorded in I/13 and was significantly superior than other TPS families. Higher average tuber weight was recorded with wider spacing (50x20 cm) and was significantly higher than other spacing treatments. Higher average tuber weight recorded in I/13 could be due to the more total tuber weight and weight of large size tubers. The more average tuber weight in wider spacing was perhaps due to the less number of tubers and higher proportion of tubers in large and medium size grades. Among the families, significantly higher tuber yield (q/ha) was recorded in II/67. The 50x15 cm spacing yielded higher tuber yield as compared to other treatments and was statistically at par with wider spacing (50x20 cm) but significantly superior than closer (50x10 cm) spacing. However, maximum marketable tuber yield recorded with TPS family I/13 was statistically at par with other families. Higher marketable tuber yield (%) was recorded with wider spacing (50x20 cm) and was significantly more than 50x10 cm spacing but at par with 50x15 cm spacing.

Table 4.6.3: Average tuber weight, tuber yield and marketable tuber yield in different TPS families and spacing (pooled over 1996-97)

Treatment	Av. tuber wt. (g)	Tuber yield (q/ha)	Marketable tuber yield (%)
TPS families			
I/13	54.0	405	89.9
II/13	46.0	394	86.5
I/67	48.5	397	88.3
II/67	48.5	454	88.1
7/67	46.0	408	87.1
CD (0.05)	3.73	37.67	3.9
Spacing (cm)			
50 x 10	44.0	384	84.8
50 x 15	47.2	453	88.1
50 x 20	54.7	422	91.1
CD (0.05)	3.66	35.80	3.0

The interaction of various TPS families and spacing treatments influenced total number of tubers and number of tubers (per plot) in different grades (Table 4.6.4). Higher total number of tubers was recorded in 7/67 and 50x15 cm (V_5S_2) treatment and was significantly higher than $V_1S_{1,}$ $V_1S_{2,}$ $V_2S_{3,}$ and V_3S_1. Higher number of tubers in large (> 40 g) size grade was recorded in II/67 and 50x15 cm (V_4S_2) treatment and was significantly at par with S_2 V_5, V_1S_3 and V_3S_2 and statisticaly superior to all other treatments. Maximum number of tubers in medium size grade was recorded in II/67 and 50x15 cm (V_4S_2) but differences were non significant among the treatments. Similarly 7/67 and 50x15 cm (V_5S_2) produced more number of tubers in small size (<20 g) grade but did not differ statistically than all other treatments.

Table 4.6.4: Interaction effect of different TPS families and spacing on total number of tubers and number of tubers in different grades (pooled over 1996-97)

Spacing (cm)	**TPS Families**					
	I/13 (V_1)	**II/13 (V_2)**	**I/67 (V_3)**	**II/67 (V_4)**	**7/67 (V_5)**	**Mean**
			(1) Total no. of tuber/plot			
50 x 10 (S_1)	274.1	337.0	131.6	339.0	336.3	283.6
50 x 15 (S_2)	283.6	324.8	332.8	362.5	384.4	337.6
50 x 20 (S_3)	304.8	288.8	304.5	339.3	339.4	315.4
Mean	287.5	316.8	256.3	346.9	353.4	
			(2) No. of tubers/plot (>40g)			
50 x 10 (S_1)	92.1	72.2	102.5	98.4	84.6	90.0
50 x 15 (S_2)	92.6	92.8	105.3	138.6	111.3	108.1
50 x 20 (S_3)	108.4	96.3	82.3	100.4	82.6	94.0
Mean	97.7	87.1	96.7	112.5	92.8	
			(3) No. of tubers/plot (20-40g)			
50 x 10 (S_1)	93.1	131.5	99.5	105.3	122.6	110.4
50 x 15 (S_2)	91.6	126.8	113.4	137.1	117.9	117.4
50 x 20 (S_3)	105.3	96.8	100.1	111.6	104.9	103.7
Mean	96.7	118.4	104.3	118.0	115.1	

contd...

table contd...

	(4) No. of tubers/plot (<20g)					
50 x 10 (S_1)	87.1	133.1	94.3	134.8	132.6	116.4
50 x 15 (S_2)	82.6	147.1	114.8	136.3	155.1	127.2
50 x 20 (S_3)	93.6	95.5	100.3	127.4	151.8	113.7
Mean	87.8	125.2	103.1	132.8	146.5	

CD (0.05)	(1)	(2)	(3)	(4)
S x F	86.1	31.4	43.1	60.64

Combined effect of TPS families and spacing treatments influenced the total weight of tubers (kg/plot) and weight of tubers in different grades (Table 4.6.5). Maximum total weight of tubers was recorded in II/67 and 50x15 cm ($V_4 S_2$) treatment but it was statistically at par with 7/67 and 50x20 cm ($V_5 S_2$) and significantly superior to all other treatments. Higher weight of tubers of large size (>40 g) grade was recorded with $V_4 S_2$ and was at par with $V_1 S_2$, $V_3 S_2$ and $V_5 S_2$ and significantly superior to other treatments. Tubers in medium size (20-40 g) grade in the treatment consisting II/13 TPS family and 50x10 cm spacing yielded significantly more tuber weight than I/13 & II/13 planted at 50x20 cm spacing but the difference was non significant over other treatments. The low yield under V_1S_3 and V_2S_3 treatments may be due to less number of plants under wider spacing than V_2S_1 treatment. Non significant difference among the treatments was recorded on weight of tubers for small (< 20 g) size tuber grade.

Table 4.6.5: Interaction of different TPS families and spacing on total weight of tubers and weight of tubers of different grades (pooled over 1996-97)

Spacing (cm)	**TPS Families**					
	I/13 (V_1)	**II/13 (V_2)**	**I/67 (V_3)**	**II/67 (V_4)**	**7/67 (V_5)**	**Mean**
	(1) Total wt. of tuber/plot					
50 x 10 (S_1)	12.5	13.0	14.5	14.6	13.4	13.6
50 x 15 (S_1)	15.5	14.6	15.5	18.7	16.2	16.1

contd...

table contd...

50 x 20 (S_1)	13.2	12.3	13.5	15.1	13.2	13.5
Mean	13.7	13.3	14.5	16.1	14.2	
			(2) Wt. of tubers/plot (>40g)			
50 x 10 (S_1)	8.6	8.0	10.6	10.3	8.5	9.2
50 x 15 (S_1)	11.3	9.8	11.6	14.5	11.3	11.7
50 x 20 (S_1)	9.5	9.2	8.8	9.6	8.5	9.1
Mean	9.8	9.0	10.3	11.5	9.4	
			(3) Wt. of tubers/plot (20-40g)			
50 x 10 (S_1)	3.3	4.8	3.1	4.2	3.6	3.8
50 x 15 (S_1)	3.5	4.4	3.6	4.6	4.2	4.1
50 x 20 (S_1)	2.8	2.9	3.4	4.4	3.8	3.5
Mean	3.2	4.0	3.4	4.4	3.9	
			(4) Wt. of tubers/plot (<20g)			
50 x 10 (S_1)	1.1	1.6	1.2	1.8	1.6	1.5
50 x 15 (S_1)	1.0	1.9	1.3	1.8	1.7	13.
50 x 20 (S_1)	1.0	1.1	1.4	1.5	1.5	1.1
Mean	1.0	1.5	1.3	1.7	1.6	
CD (0.05)	(1)		(2)		(3)	(4)
S x F	2.2		3.4		1.7	1.3

It is clear from Table 4.6.6 that the interaction of TPS families and spacing influenced the average tuber weight, tuber yield and marketable tuber yield percentage. The average tuber weight of the TPS families increased with increasing plant to plant spacing. However, maximum average tuber weight was recorded in I/13 and 52x20 cm (V_1 S_3) treatment and was significantly higher than other treatments. Higher average tuber weight in V_1S_3 treatment might be due the genotype of I/13 (V_1) producing comparatively large size tubers with wider spacing. The maximum yield of tubers (q/ha) of all the families were recorded with medium spacing (50x15 cm). Maximum tuber yield (495 q/ha) was recorded with the combination of II/67 (V_4) family and 50 x 15 cm (S_2) spacing. The treatment V_4S_2 produced more tuber yield (q/ha) than all other treatments and the difference was significantly higher than V_1S_1, V_2S_1,

V_2S_3, V_5S_1, V_5S_3 and V_4 S_1. Marketable tuber yield increased with increasing spacing. Wider spacing (50x20 cm) resulted in significantly higher yield than closer (50x10 cm) spacing in case of II/13 and I/67. However, the marketable tuber yield of all the families and medium spacing (50 x 15 cm) was at par with wider spacing treatment.

Table 4.6.6: Interaction of different TPS families and spacing on average tuber weight, tuber yield and marketable tuber yield (pooled over 1996-97).

Spacing (cm)	TPS Families					
	I/13 (V_1)	II/13 (V_2)	I/67 (V_3)	II/67 (V_4)	7/67 (V_5)	Mean
	(1) Av. per tuber wt. (g)					
50 x 10 (S_1)	45.9	42.0	47.1	44.9	39.2	43.8
50 x 15 (S_2)	51.5	44.2	48.4	47.2	44.8	47.2
50 x 20 (S_3)	60.9	52.1	53.1	53.4	49.3	53.8
Mean	52.8	46.1	50.0	48.5	44.4	
	(2) Tuber yield (q/ha)					
50 x 10 (S_1)	335	382	423	391	392	384
50 x 15 (S_2)	446	424	483	495	436	457
50 x 20 (S_3)	434	378	421	482	395	422
Mean	405	394	442	456	408	
	(3) Marketable tuber yield (%)					
50 x 10 (S_1)	88.5	82.5	82.5	86.8	83.7	84.8
50 x 15 (S_2)	90.1	84.2	90.0	87.8	87.8	88.0
50 x 20 (S_3)	91.3	92.9	92.4	89.9	89.8	91.3
Mean	90.0	86.5	88.3	88.2	87.1	
CD (0.05)	(1)		(2)			(3)
S x S	5.2		70.2			6.0

Effect of Seedling Tuber Size and Plant Spacing on Potato Production

Proper seed size and optimum plant density have a great influence on the growth of potato crop, tuber yield and economic returns. Due to long photoperiod and crop duration, the ware potato crop produces large size tubers at normal plant population (67000 plants/ha). If the large size seed tubers are used for planting at normal plant density, the seed rate is increased hence not economical for potato production. To determine the effect of plant spacing and size of seedling tubers on yield attributes, the present study was carried out in a factorial randomized block design with three replications. Treatment comprised of combination of four seedling tuber size 10, 20, 30 and 40 g and three levels of plant spacing 45x10, 45x15, 45x20 cm. The tubers of II/67 hybrid TPS family produced during previous season and stored with underground storage method in winter were planted in first week of May during 1998 and 1999. Cultural practices were followed as described in the earlier chapter. The crop was dehaulmed in second week of September (125 DAP) and harvested after another 15 days. The produce of the plots size (1.80x1.80 m) was graded and tuber numbers and weight of each grades >40g (large), 20-40g (medium) and <20g (small) were recorded. More than 20 g tubers were considered as marketable tuber yield. Both the years data was analysed on pooled basis.

Data presented in Table 4.6.7 indicate although non significant difference in total number of tubers with different seedling tuber size treatments yet the number of total tubers increased with decreasing plant spacing. More number of total tubers was recorded at 45x10 cm spacing which were significantly higher over 45x20 cm but at par with 45x15 cm spacing. The number of tubers/m^2 increased with increase in plant density have also been reported by Singh *et al.* (1997). Number of large size tubers increased with increases size of seedling tuber. However, higher number of tubers of >40 g size grade was recorded at closer spacing (45x10 cm) and the lowest with wider spacing (45x20 cm). The difference between number of tubers under medium (20-40 g) and small (< 20 g) size tubers grade were found to be non significant with increase in seedling tuber size and spacing.

The data (Table 4.6.8) revealed a significant difference in total tuber weight and weight of tubers under large size grades with different seedling tuber size and spacing treatments. The highest total weight of tubers was recorded when large size (40 g) seedling tubers were planted at 50x20 cm spacing. In case of medium (20-40 g) and small (<20 g) size tuber grades, there was no significant difference for response recorded among all seedling tuber size and spacing treatments. Roy and Srivastwa (1996) have also been reported higher potato yield with 50 g seed size tubers. These findings are also in agreement with the results of Nandekar (2005) and Kadian *et al.* (1988).

Table 4.6.7: Performance of total and number of tubers in different grades with various seedling tubers size and spacing.

Treatment	Total no. of tubers/plot	No. of tubers/plot (>40g)	No. of tubers/plot (20-40g)	No. of tubers/plot (<20g)
Seedling tuber size (g)				
10	256.5	120.9	76.0	59.4
20	256.4	110.8	76.7	63.2
30	274.5	142.5	69.9	54.5
40	268.6	156.6	74.2	51.7
CD (0.05)	37.8	31.0	18.0	13.5
Spacing (cm)				
45 x 10	280.6	143.3	79.5	58.1
45 x 15	274.7	134.6	79.1	60.8
45 x 20	236.8	120.2	63.9	52.9
CD (0.05)	22.7	27.0	15.3	10.4

Table 4.6.8: Performance of total weight and weight of tubers under different grades with various seedling tubers size and spacing.

Treatment	Total wt. of tubers kg/plot	Wt. of tubers kg/plot (>40g)	Wt. of tubers kg/plot (20-40g)	Wt. of tubers kg/plot (<20g)
Seedling tuber size (g)				
10	11.0	8.5	1.7	0.85
20	13.4	10.1	2.2	1.05
30	15.3	11.9	2.0	1.45
40	16.9	14.0	1.8	1.00
CD (0.05)	0.68	0.37	0.5	0.37
Spacing (cm)				
45 x 10	13.1	10.1	1.8	1.1
45 x 15	14.3	11.2	1.8	1.2
45 x 20	15.3	12.0	2.0	1.0
CD (0.05)	0.58	0.34	0.4	0.32

Average tuber weight, tuber yield and marketable tuber yield were greatly influenced by different seedling tuber size and spacing treatments (Table 4.6.9). Progressive increase in average tuber weight was obtained with increase in seedling tuber size and plant to plant spacing. Significantly higher tuber weight with large size (40 g) seedling tubers recorded which may be due to the higher photosynthetate translocation compared to other treatments. Production of large size tubers with 50 g seed size tuber has also been reported by Nandekar (2005). The yield of large size tubers increased with decreasing seed rate (Singh, 1997).

In this Study, tuber yield (q/ha) increased with increasing seedling tuber size and plant to plant spacing. Maximum tuber yield (422 q/ha) was recorded with large size (40 g) seedling tubers and was significantly higher than the other treatments. Similar trend was observed with spacing and more tuber yield

(344 q/ha) recorded at wider (45x20 cm) spacing and was significantly superior to 45x10 cm and 45x15 cm spacing. The favorable effect of 45x20 cm spacing observed might be due to the wide spacing which produce more number of tubers (per plot), weight and large size of the tubers. Similar results have also been reported by Ahire *et al.* (2000).

Table 4.6.9. Average tuber weight, tuber yield and marketable tuber yield as influence by different seedling tuber size and spacing

Treatment	Av. per tuber wt. (g)	Tuber yield (q/ha)	Marketable tuber yield (%)
Seedling tuber			
size (g)			
10	42.72	235	78.6
20	44.26	270	81.2
30	52.97	300	79.8
40	68.68	422	86.6
CD (0.05)	5.83	47	3.8
Spacing (cm)			
45 x 10	44.51	271	75.6
45 x 15	51.98	305	81.7
45 x 20	59.98	344	88.1
CD (0.05)	2.94	23	2.4

In this experiment, maximum marketable tuber yield was recorded with 40g seedling tuber size (86.6%) and was statistically higher over other tuber size. Higher marketable tuber yield was recorded with 45x20 cm (88.1%) spacing than 45x10 cm (75.6%) and 45x15 cm (81.7%). The higher marketable tuber yield with large size seedling tubers and wider spacing could be due to production of more number and weight of tubers from large size grade. These results are in accordance with Singh (1997) and Wurr *et al.* (1992).

Table 4.6.10. Interaction effect of different seedling tuber size and spacing on total number of tuber and number of tubers under different grade

Spacing (cm)	Seedling tuber size (g)				
	10	**20**	**30**	**40**	**Mean**
	(1) Total no. of tubers / plot				
45 x 10	283.1	275.6	264.0	299.9	280.7
45 x 15	254.5	264.6	260.1	276.4	274.6
45 x 20	231.9	229.3	256.4	229.5	236.8
Mean	256.5	180.1	274.5	268.6	
	(2) No. of tubers/plot (>40g)				
45 x 10	141.8	127.1	142.6	161.8	143.3
45 x 15	111.1	118.3	149.1	139.8	134.6
45 x 20	109.8	86.9	157.9	116.1	120.2
Mean	121.0	110.8	156.5	142.6	
	(3) No. of tubers/plot (20-40g)				
45 x 10	83.9	85.1	71.9	77.1	79.5
45 x 15	77.9	81.1	60.3	87.3	76.7
45 x 20	66.3	63.8	67.4	58.3	64.0
Mean	76.0	76.7	66.5	74.2	
	(4) No. of tubers/plot (<20g)				
45 x 10	57.3	63.0	50.9	60.9	58.0
45 x 15	65.3	65.1	63.6	49.3	60.8
45 x 20	55.8	61.8	49.1	44.9	52.5
Mean	59.5	63.3	54.5	51.7	

CD (0.05)	(1)	(2)	(3)	(4)
S x W	55.5	33.8	31.1	32.2

NS - non significant

Total tuber number and number of tubers under large size tuber grade increased with decrease in inter row spacing (Table 4.6.10). More number of tubers was recorded where large size (40g) seedling tubers were planted at closer spacing (45x10 cm). The non significant difference in total number of tubers and number of tubers under large size grade were recorded at wider (45x20 cm) and closer (45x10 cm) spacing. Although higher number of tubers in 20-40 g tuber size grade was recorded under 40 g size seedling tuber planted at 45x15 cm and in < 20 g tuber grade with 10 g size seedling tuber at 45x15 cm but significant effect of seedling tuber size and spacing treatments was not obtained on number of tubers from medium (20–40 g) and small (< 20 g) grades.

The data presented in Table 4.6.11 revealed that the total weight of tubers (per plot) increased with increase in seedling tuber size in all the spacing treatments. Maximum total weight of tubers (17.9 kg/plot) was recorded with 40 g seedling tubers planted at 45x20 cm spacing. This may be due to the bigger size (40 g) seedling tuber produces more number of tubers with all the spacings but wider spacing offers more nutrients to lesser number of plants which resulted production of comparatively bigger size tubers. Maximum tuber weight of large size tuber (>40 g) grade was observed with large size seedling tuber (40 g) and medium spacing (45x15 cm) and was at par with wider spacing (45x20 cm) and significantly higher than other treatments. Maximum tuber weight in 20-40g grade was recorded with 20 g seedling tuber planted at 45x15 cm. More tubers weight in small size (<20 g) grade was recorded with 40 g seedling tuber planted at 45x20 cm spacing and was significantly higher over other treatments. The higher weight of tubers was due to production of larger size tuber when planted at wider spacing.

Table 4.6.11: Interaction effect of different seedling tuber size and spacing on total wt. of tubers and wt. of tuber under different grade

Spacing (cm)	Seedling tuber size (g)				
	10	**20**	**30**	**40**	**Mean**
	(1) Total wt. of tuber/plot				
45 x 10	10.0	12.3	13.8	16.1	13.1
45 x 15	11.0	13.4	15.8	16.1	11.6
45 x 20	12.2	14.3	15.7	17.9	15.0
Mean	11.1	13.3	15.1	16.7	

contd...

table contd...

	(2) Wt. of tubers/plot (>40g)				
45 x 10	7.7	9.1	10.9	13.0	10.2
45 x 15	8.4	10.2	12.1	15.0	11.3
45 x 20	9.4	11.1	12.7	14.3	12.1
Mean	8.5	10.1	11.9	14.1	
	(3) Wt. of tubers/plot (20-40g)				
45 x 10	1.5	2.2	2.0	1.9	1.9
45 x 15	0.8	2.3	2.1	1.5	1.7
45 x 20	1.9	2.1	1.8	2.1	2.0
Mean	1.4	2.2	2.0	1.9	
	(4) Wt. of tubers/plot (<20g)				
45 x 10	0.8	1.1	1.2	1.1	1.1
45 x 15	1.0	1.0	1.1	1.0	1.0
45 x 20	1.0	1.2	1.3	2.0	1.4
Mean	1.1	1.2	1.2	1.3	

CD (0.05)	(1)	(2)	(3)	(4)
S x W	1.4	0.7	0.9	0.5

It is evident from Table 5.6.12 that maximum average tuber weight, tuber yield and marketable tuber yield were produced with seedling tubers of 40 g planted at wider (45x20 cm) spacing. Therefore, significantly the highest average tuber weight (76.7 g), tuber yield (476 q/ha) and marketable tuber yield (92.9%) were recorded with 40 g seedling tuber planted at 45x20 cm The bigger size seedling tuber planted at wider spacing increased the size of tubers, weight of tubers per plant over closer spacing. Ahire *et al.,* (2000) and Singh (1983) also reported higher weight of tubers in single row planting than double row planting. Mahmood and Gill (1984) however reported higher tuber weight and yield with 30-50 g seed tubers.

Table 4.6.12: Interaction effect of different seedling tuber size and spacing on average tuber weight, tuber yield and marketable tuber yield.

Spacing (cm)	Seedling tuber size (g)				
	10	**20**	**30**	**40**	**Mean**
		(1) Av. per tuber wt. (g)			
45 x 10	35.9	34.6	47.3	60.0	44.5
45 x 15	44.8	42.4	51.7	69.0	52.0
45 x 20	53.5	49.8	59.9	76.7	60.0
Mean	44.7	42.3	53.0	68.7	
		(2) Tuber yield (q/ha)			
45 x 10	206	255	237	388	271
45 x 15	240	255	320	403	305
45 x 20	260	301	288	476	331
Mean	235	271	282	422	
		(3) Marketable tuber yield (%)			
45 x 10	73.0	74.9	75.1	80.9	76.0
45 x 15	78.5	82.5	78.3	87.1	81.6
45 x 20	84.4	89.1	85.8	92.9	88.0
Mean	78.6	82.2	79.7	87.0	
CD (0.05)	(1)		(2)		(3)
S x W	5.8		46.5		4.8

It can be concluded from the study that all size seedling tubers of TPS family II/67 have potential to raise potato production by planting them at different spacing but maximum yield can be obtained with 40 g seedling tubers planted at 45x20 cm spacing in high altitude cold desert region.

❑❑❑

CHAPTER **5**

Development & Evaluation of New TPS Families

1

2

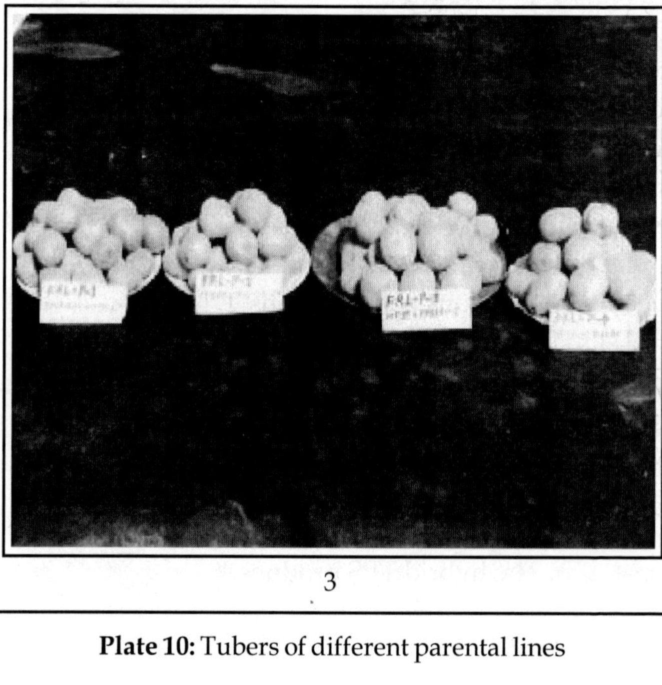

3

Plate 10: Tubers of different parental lines

Effect of Planting Spacing of locally developed TPS Progenies Raised through Seedling Tuber on yield

In the field survey, potato germplasm was collected on the basis of variation in morphological characters of plant and tubers from far–flung villages of Ladakh region. The germplasm was evaluated at the experimental farm of the Institutes out of six, two lines were identified as andigena and other 4 as tuberousum species. One suitable line from each group was used in hybridization with the TPS lines procured from International Potato Centre (CIP) SWA Region, New Delhi. Total 12 crosses were attempted and four best combinations were selected for further study. This investigation was carried out to grow selected progenies through seedling tubers planted at different spacing with the aim to find out suitable progeny to make the region self sufficient in quality planting material for higher yield and quality attributes.

Two parental lines TPS-FRL-1 and TPS-FRL-2 were selected from the germplasm collected from different locations of Ladakh region. Two parental lines TPS-67 (male) and MF-II (female) were used from the lines provided by CIP. 30-40 g tubers were selected from these four lines and stored in winter season. The selected tubers were planted in separate blocks in the month of May in next year. Hybrid TPS was produced from the four combinations TPS-FRL-1xTPS-67 (P_1), TPS-FRL-2xTPS-67 (P_2), MF-II x TPS FRL-2 (P_3) and MF-II x TPS-FRL-1 (P_4). To raise the nursery, the TPS produced from all the four progenies was sown next year in greenhouse in first week of April. The seedlings were transplanted in last week of May as per standard package of practices. From the crop 30-40 g size seedling tubers were selected and stored for the winter season. The hybrid TPS produced during 1998 was used for subsequent two years for this investigation. The selected seedling tubers of all the four progenies were planted in second week of May in both the years at 45x10 cm (S_1), 45x15 cm (S_2) and 45x20 cm (S_3) in plots measuring 1.80x1.80 m in randomized block design with three replications. All the standard cultural practices were followed and the crop was harvested in last week of September. The produce was graded in three grades *i.e.* >40 g, 20–40 g, and <20 g. The number and weight of the tubers in all three grades were recorded. The tuber yield (g/ha) calculated leaving 10 percent area for layout of the field. More than 20 g tubers were considered as marketable yield. Both the years' data were analyzed on pool basis.

Number of tubers (per plot) was significantly influenced by TPS progenies and spacing (Table 5.1). Higher number of total tubers and >40 g grade tubers were recorded in P_3 treatment and were significantly higher than the other progenies (P_1, P_2 and P_4). Under 20-40 g grade more number of tubers was recorded in P_3 progeny which was at par with P_2 but significantly higher than the other progenies. Significant difference in number of tubers of the progenies was not obtained under < 20 g grades. The difference in number of tubers (per plant) between the cross combination of different accessions of potato has also been reported by Bhardwaj *et al.* (2007). In this experiment the spacing treatments had no significant effect on the total number of tubers (per plot). But the number of large size tubers increased with increcising spacing and significantly higher number of tubers under >40 g and 20-40 g grades recorded at wider spacing (45x20 cm). Small size tuber grade (<20 g) were significantly lower at the wider spacing (45x20 cm) and higher number of tubers from closer spacing (45x10 cm). The planting distance of 45x20 cm for 20-40 g seedling tubers size has also been recommended in "Manual for TPS production and utilization, CPRI, Shimla".

Table 5.1: Effect of TPS progenies and spacing on total number of tubers and number of tubers in different grade / plot

Treatment	Total no. of tubers /plot	No of tubers/ plot (>40g)	No. of tubers/ plot (20-40g)	No. of tubers/ plot (<20g)
Progeny				
TPS-FRL-1 x TPS-67 (P_1)	238.2	75.9	90.9	85.82
TPS-FRL-2 x TPS-67 (P_2)	269.9	55.5	106.3	109.5
MF-II x TPS-FRL-2 (P_3)	322.9	89.8	115.3	109.3
MF-II x TPS-FRL-1 (P_4)	267.8	77.0	88.9	105.3
CD (0.05)	34.8	4.1	9.7	14.2
Spacing				
45 x 10 (S_1)	265.9	60.2	83.3	130.6
45 x 15 (S_2)	272.6	66.9	110.6	114.9
45 x 20 (S_3)	283.6	77.9	115.6	89.8
CD (0.05)	26.5	2.6	2.7	7.2

Maximum weight of total tubers and tubers in grades of >40 g and 20-40 g was recorded in P_3 treatment followed by P_4 and was significantly higher than the other treatments (Table 5.2). Among the TPS progenies maximum weight of tubers was recorded in P_1 progeny but no significant variation observed in weight of small size (<20 g) grade. Higher weight in >40g and 20-40 g grades recorded was mainly due to more number of bigger size tubers obtained in these grades. Difference in average tuber weight with various TPS progenies have also been reported by Pandey *et al.* (2003). Significantly higher weight of total tubers was recorded at wider spacing (45x20 cm). Similar trend was also observed in >40 g and 20-40 g grades. However, with <20 g tubers grade, significantly more tuber weight was recorded at 45x15 cm spacing. The higher tuber weight in large and medium tuber size grade at 45x20 cm spacing was on account of more large size tubers and the results are in line with those obtained by Dubey (2003).

Table 5.2. Weight of tubers/plot as influenced by various TPS progenies and different spacing

Treatment	Total wt. of tubers kg/plot	Wt. of tubers kg/plot (>40g)	Wt. of tubers kg/plot (20-40g)	Wt. of tubers kg/plot (<20g)
Progeny				
TPS-FRL-1 x TPS-67 (P_1)	10.20	5.35	3.34	1.71
TPS-FRL-2 x TPS-67 (P_2)	9.30	4.53	2.68	1.60
MF-II x TPS-FRL-2 (P_3)	13.10	7.37	4.09	1.66
MF-II x TPS-FRL-1 (P_4)	11.01	6.43	3.22	1.38
CD (0.05)	0.61	0.59	0.57	0.45
Spacing				
45 x 10 (S_1)	8.95	4.89	2.54	1.49
45 x 15 (S_2)	10.81	5.91	3.22	1.57
45 x 20 (S_3)	12.02	6.96	4.24	1.37
CD (0.05)	0.31	0.30	0.35	0.23

Different TPS progenies and spacing treatments had effect on the total yield and marketable tuber yield (Table 5.3). Data showed that maximum tuber yield 353 q/ha was produced by P_3 followed 303 q/ha in P_4. In respect of total tuber yield, P_3 was superior than the other treatments. More tuber yield in P_3 might be attributed due to the more number and weight of the tubers. Higher tuber yield with large size tubers in potato have also been reported by Suman *et al.,* (2003). Inter-relationship between genetic parameters for tuber yield in potato have also been reported by Luthra *et al,* (2001) and good general combining ability for yield/plot by Kumar and Gopal (2003) in certain accessions of Andigena. Tuber yield at wider & medium spacing was significantly higher than close spacing but was at par with medium spacing.

Table 5.3: Effect of different TPS progenies and spacing on tuber yield and % marketable yield

Treatment	Tuber yield (q/ha)	Marketable yield (%)
Progeny		
TPS-FRL-1 x TPS-67 (P_1)	289	90.8
TPS-FRL-2 x TPS-67 (P_2)	244	93.3
MF-II x TPS-FRL-2 (P_3)	353	92.0
MF-II x TPS-FRL-1 (P_4)	303	88.7
CD (0.05)	44.34	7.3
Spacing		
45 x 10 (S_1)	255	85.3
45 x 15 (S_2)	300	87.8
45 x 20 (S_3)	324	93.0
CD (0.05)	24.97	3.76

No significant variation was recorded among the progenies in marketable yield but significantly higher marketable yield percentage was recorded at wider spacing (45x20 cm). Whereas, no significant variation in marketable yield percentage was obtained with close (45x10 cm) and medium spacing (45x15 cm). Higher marketable yield at 45x20 cm spacing could be due to the increased number of large size tubers with increase in plant spacing. These results are in agreement of the findings of Patel *et al.* (2003) who have reported increase in number of large size tuber with increased plant spacing in potato crop.

The interaction effect of hybrid TPS progenies and spacing greatly influenced the total number of tubers and tubers number under different grades (Table 5.4). The total number of tubers (per plot) ranged between 207 & 311 significantly the highest total number of tubers were recorded with $P_3 S_3$ treatment over $P_1 S_1$, $P_1 S_2$ & $P_4 S_3$ while difference was non-significant with other treatments. Number of tubers ranged from 49.6-90.2 under > 40 g grade and significantly higher number of tubers was recorded with $P_3 S_3$ which also showed higher number of tubers under 20-40 g grade. The progeny MF-IIxTPS-FRL-2 produced more total number of tubers at wider spacing with large (>40 g) and medium (20-40 g) grades.

Table 5.4: Interaction effect of different progenies and spacings on number of tubers

Spacing (cm)	Progenies				
	TPS-FRL-1 x TPS-67 (P_1)	TPS-FRL-2 x TPS-67 (P_2)	MF-II x TPS-FRL-2 (P_3)	MF-II x TPS-FRL-1 (P_4)	Mean
	(1) Total no. of tubers / plot				
45 x 10 (S_1)	207	285	292	281	266
45 x 15 (S_2)	232	278	275	306	273
45 x 20 (S_3)	277	280	311	267	284
Mean	238	281	293	285	
	(2) No. of tubers/plot (>40g)				
45 x 10 (S_1)	63.0	49.6	52.6	75.3	60.1
45 x 15 (S_2)	73.1	54.5	66.6	73.5	66.9
45 x 20 (S_3)	81.6	62.5	90.2	82.1	79.1
Mean	72.6	55.5	69.8	77.0	
	(3) No. of tubers/plot (20-40g)				
45 x 10 (S_1)	65.2	88.8	100.0	79.7	83.4
45 x 15 (S_2)	92.3	98.0	114.5	95.1	100.0
45 x 20 (S_3)	110.5	108.0	131.6	92.1	110.6
Mean	89.3	98.3	115.4	89.0	
	(4) No. of tubers/plot (<20g)				
45 x 10 (S_1)	106.5	143.0	145.7	127.2	130.6
45 x 15 (S_2)	81.5	121.0	150.1	107.0	115.0
45 x 20 (S_3)	79.5	97.7	95.1	82.0	88.6
Mean	89.2	120.5	130.3	105.4	

CD (0.05)	(1)	(2)	(3)	(4)
S x P	52.0	5.1	11.5	14.4

Total tubers weight and weight of different grades tubers (per plot) was affected by the interaction of different TPS progenies and spacings (Table 5.5). The total weight of tubers ranged from 8.0 kg to 14.4 kg/plot and significantly higher total tuber weight was recorded in $P_3 S_3$. Similar results were obtained with >40g and 20-40 g grades where significantly higher yield (8.40 and 5.5 kg/plot) was recorded with $P_3 S_3$ treatment. However, no significant difference in tuber yield was obtained for < 20 g grade tubers. The progeny MF-II x TPS-FRL-2 produced more total weight and weight of tubers in the large and medium grades at wider spacing could be due to potential of the parental line to produce large size tubers at wider spacing.

Table 5.5: Interaction of different progenies and spacings on weight of tubers

Spacing (cm)	Progenies				
	TPS-FRL-1 x TPS-67 (P_1)	TPS-FRL-2 x TPS-67 (P_2)	MF-II x TPS-FRL-2 (P_3)	MF-II x TPS-FRL-1 (P_4)	Mean
	(1) Total wt. of tubers/plot				
45 x 10 (S_1)	8.6	8.0	10.4	9.0	9.0
45 x 15 (S_2)	10.2	9.0	12.1	11.1	10.6
45 x 20 (S_3)	11.9	9.4	14.4	13.0	12.2
Mean	10.2	8.8	12.3	11.0	
	(2) Wt. of tubers/plot (>40g)				
45 x 10 (S_1)	4.4	3.5	6.4	5.4	4.9
45 x 15 (S_2)	5.3	4.5	7.4	6.5	5.9
45 x 20 (S_3)	6.5	5.5	8.4	7.5	7.0
Mean	5.4	4.5	7.4	6.5	
	(3) Wt. of tubers/plot (20-40g)				
45 x 10 (S_1)	2.6	2.3	2.8	2.5	2.6
45 x 15 (S_2)	3.3	2.8	4.0	2.7	3.2
45 x 20 (S_3)	4.1	2.9	5.5	4.5	4.3
Mean	3.3	2.7	4.1	3.2	

contd...

table contd...

	(4) Wt. of tubers/plot (<20g)				
45 x 10 (S_1)	1.5	1.4	1.1	1.5	1.4
45 x 15 (S_2)	1.7	1.2	1.5	1.3	1.4
45 x 20 (S_3)	1.8	1.2	1.1	1.4	1.3
Mean	1.7	1.3	1.2	1.4	
CD (0.05)	(1)	(2)	(3)	(4)	
P x S	0.6	0.5	0.6	0.5	

Table 5.6 showed that the total yield and % marketable tuber yield varied with hybrid TPS progeny and spacing treatments. Maximum tuber yield was recorded in P_3S_3 (370 q/ha) and was at par with P_3S_2 (349 q/ha) and P_4S_3 (342 q/ha) and significantly superior to the other treatments. Higher marketable tuber yield was recorded in P_3S_3 but was statistically at par with P_4S_3, P_3S_2 and P_1S_3 and significantly higher than the other treatments. The progeny MF-IIxTPS-FRL-2 with wider spacing (45x20 cm) produced more large size tubers of >40 g and 20-40 g grade could be reason for higher marketable tuber yield under this treatment.

Table 5.6: Interaction of different progenies and spacings on tuber yield and marketable tuber yield percentage

Spacing (cm)	Progenies TPS-FRL-1 x TPS-67 (P_1)	TPS-FRL-2 x TPS-67 (P_2)	MF-II x TPS-FRL-2 (P_3)	MF-II x TPS-FRL-1 (P_4)	Mean
	(1) Tuber yield (q/ha)				
45 x 10	246	230	302	243	255
45 x 15	86	242	349	309	296
45 x 20	320	261	370	342	323
Mean	284	244	340	298	
	(2) Marketable tuber yield (%)				
45 x 10	84.6	77.5	90.8	84.8	84.4
45 x 15	89.7	85.7	93.3	74.9	85.9

contd...

table contd...

45 x 20	91.9	86.8	97.4	96.4	93.1
Mean	88.7	83.3	93.8	85.4	
CD (0.05)		(1)		(2)	
S x P		30.3		5.3	

It can be concluded from the above discussion that 30-40 g seedling tubers of progeny MF-IIxTPS-FRL-2 planted at 45x20 cm spacing give more total tuber yield and number of tubers under large and medium size tuber grades hence, has higher percent marketable tuber yield in conditions of cold desert.

□□□

CHAPTER 6

Mulching Effect on Productivity

Mulching in potato crop

Growth of potato crop with mulching

Plate 11

Role of Polythene Mulching on Growth and Yield of Potato

Snow melted water is the main source for irrigation in the region, which is available in limited quantity due to the physiographical constraint. Therefore, innovative methods of potato cultivation are required for indicious application of water. Being a desert, the region also has low humidity during cropping season, which leads to high evapotranspiration rate therefore need more number of irrigations. Polythene mulching not only helps to conserve soil moisture but also increases the soil temperature (Aulakh and Sur, 1999 and Bajwa *et al.* 1970). Low soil temperature specially at initial stage of growth is a major constraint for getting low yield. Keeping this in view, an investigation was carried out to determine the effect of mulching on production potential of different varieties of potato in high altitude cold desert.

Six treatment combinations consisting of no mulching, mulching with white polythene and black polythene sheets of 200 micron thickness and two cultivars Kufri Jyoti and Kufri Chandramukhi were tested. Seed tubers (size 45 g) were planted at 60 x 20 cm distance in plots measuring 2.40 x 2.40 m during 2002 and 2003 on 5th and 10th May, respectively. Organic manure @ 400 q/ha and 100:80:60 N, P_2 O_5 and K_2O /ha were applied to the soil at last ploughing. Six irrigations were given to no mulching treatment, whereas, four were given in case of white and black polythene mulching by flood irrigation method. Two hand weedings and earthing up were carried out in no mulching treatment. Haulms were cut in second week of September and crop was harvested in first week of October in both the years. Soil temperature at 15 cm depth was recorded weekly upto the 3 months. The plant height was recorded after 30 days of planting and 120 days (at haulms cutting stage). Number of stems per hill was recorded after 120 days of planting. Number and weight of total tubers per plot, average tuber weight and marketable yield (>20 g tuber) were recorded at time of harvesting.

Emergence percentage, plant height and number of stems per hill showed significant increase for all the mulching treatments over the control (Table 6.1). Plant emergence and plant height was significantly increase in polythene mulching over control with maximum emergence (97.1%), plant height at 30 DAP (25.2 cm) and at haulms cutting (99.6 cm) recorded with black polythene mulching. Kufri Chandramukhi had significantly higher emergence (92.6%) and plant height at both 30 DAP (19.6 cm) and at haulms cutting stage (91.4 cm) than Kufri Jyoti. The highest emergence and plant height obtained with black polythene mulching may be attributed to more favorable soil temperature (Fig. 6.1). Black polythene mulching treatment raised soil temperature by 5.6^0 C while white polythene mulching increased it by 2.5°C. Aulakh & Sur (1999) have reported similar rise in soil temperature with polythene mulching. This increased soil temperature helped higher emergence and fast growth of the plants. The results are in the agreement with the earlier findings (Patel *et al.* 1999; Sale, 1979). More height in Kufri Chandramukhi was due to its genetic makeup for fast and better growth under high altitude cold arid climatic conditions in comparison with Kufri Jyoti.

Table 6.1: Effect of different mulching and potato cultivars on growth and yield parameters (pooled over 2002-03)

Treatments	Emergence (%)	Plant height at 30 DAP (cm)	Plant height at haulms cutting (cm)	No. of stem/hill
Mulching				
Control (no mulch)	83.1	8.2	75.1	2.7
White polythene	92.4	19.1	85.0	5.1
Black polythene	97.1	25.2	99.6	7.4
C. D. (0.05)	0.9	1.2	2.1	0.4
Variety				
Kufri Jyoti	89.2	15.5	82.4	4.6
Kufri Chandramukhi	92.6	19.6	91.4	5.2
C. D. (0.05)	0.8	0.9	1.6	1.2

Number of stems per hill at haulms cutting stage was significantly higher with mulching (Table 6.1). Maximum number of stems per hill was recorded with black polythene mulching (7.4) and was significantly higher than those obtained with white polythene mulching (5.1) and control (2.7). Two varieties do not differ much with regard to the number of stems per hill. Black polythene mulching had more favorable temperature, which resulted in more number of stems per hill. Patel *et al.* (1999) and Midmore (1984) also reported possitive corelation with polythene mulching and number of stems per hill. Low soil temperature observed with white polythene mulching may be attributed to the fact that white polythene allowed transmission of light on soil surface. More heat loss was also recorded during night time in high altitude cold arid regions with white polythene than black polythene (Singh and Singh, 2003). Kumar *et al.* (2006) also reported wide fluctuations in diurnal soil temperature with mulching.

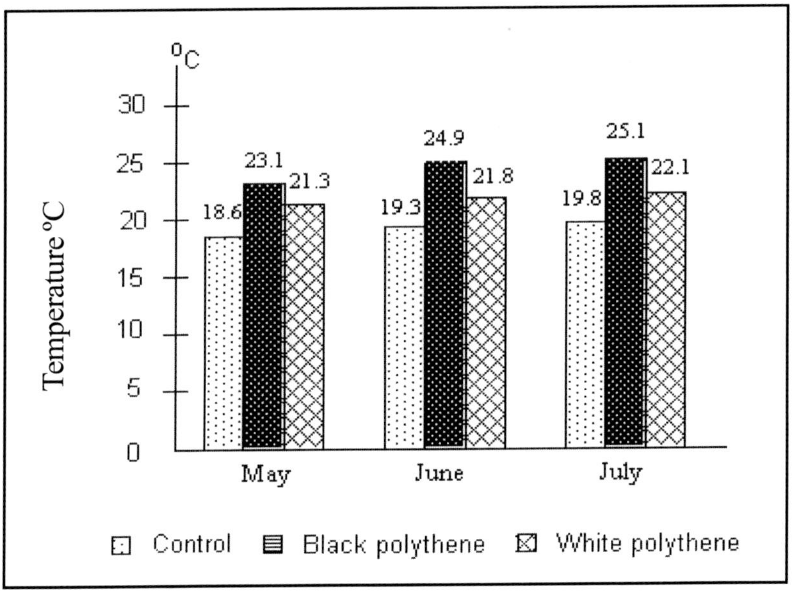

Fig. 6.1. Soil temperature at 15 cm depth (Pool over 2002-03)

Significant differences in total number of tubers, total tuber weight per plot, average tuber weight, tuber yield (q/ha) and percentage of marketable yield were observed with different mulching and the two varieties (Table 6.2). The highest number of tubers were recorded with black polythene mulching (223) and the lowest with white polythene mulching (121). Kufri Jyoti yielded significantly more number of tubers (223) than Kufri Chandramukhi (216). Higher number of tubers recorded under black polythene mulching may be due to congenial soil temperature (Fig. 6.1), which resulted in better tuberization (Yamaguchi *et al.*, 1964). The results are in agreement with relationship of number of tubers per plant and size of tubers as affected by temperature (Patel *et al.*, 2002). Higher total tuber weight (22.4 kg per plot) was recorded with black polythene mulching followed by white polythene mulching and control treatments which yielded 20.2 and 17.0 kg per plot, respectively. Kufri Chandramukhi had significantly higher tuber weight than Kufri Jyoti. Superiority of black polythene sheet as surface mulch has also been reported in tomato (Kumar *et al.*, 2006) and in Okra (Sannigrahi and Borah, 2002). Similar trend was obtained in case of average per tuber weight. The black polythene mulching treatments favors high temperature for better emergence and good growth resulting in bigger tuber size (Midmore, 1984 and Patel *et al.*, 1999).

Table 6.2: Effect of different mulching and potato cultivars on growth and yield parameters (pooled over 2002-03)

Treatments	Total number of tuber (per plot)	Total tuber weight (kg/plot)	Average per tuber wt (g)	Tuber yield (q/ha)	Marketable yield (%)
Mulching					
Control (no mulch)	214	17.0	79.5	266	85.40
White polythene	121	20.2	91.5	315	93.36
Black polythene	223	22.4	101.3	352	96.47
C. D. (0.05)	2.2	0.3	1.6	5.0	0.65

contd...

table contd...

Variety					
Kufri Jyoti	223	19.0	85.5	297	89.92
Kufri Chandramukhi	216	20.8	96.0	325	93.57
C. D. (0.05)	1.8	0.2	1.2	4.0	5.6

Significantly the highest tuber yield (352 q/ha) was recorded with black polythene mulching (315 q/ha) than obtained with white polythene mulching in case of control (266 q/ha). Higher tuber yield 325 q/ha of Kufri Chandramukhi showed significant superiority over Kufri Jyoti which recorded 297 q/ha. Maximum tuber yield with black polythene mulching might be due to higher number of tubers and per tuber weight in comparison to other treatments. Likewise, marketable yield was also higher with black polythene mulching and was statistically at par with white polythene mulching and both were significantly superior over the control. The varieties however, did not differ statistically for marketable yield.

The interaction effect of mulching and varieties was significant (Table 6.3). Black polythene mulching in Kufri Chandramukhi showed significant superiority over the other treatments in all growth and yield parameters. Thus, it can be concluded from the study that the use of black polythene mulching saves 33.3% water requirement in potato crop. Black polythene mulching also control the weeds and provide more favourable soil temperature for better emergence, growth and tuberization. The Kufri Chandramukhi has potential to produce high yield. Therefore, maximum yield of potato can be obtained by growing the cultivar Kufri Chandramukhi with black polythene mulching under high altitude cold arid climatic conditions.

... table 6.3

Table 6.3: Interaction of mulching and cultivars on potato growth and yield attributes of potato crop (pooled over 2002-03)

Treatments	Emergence (%)	Plant height at 30 DAP (cm)	Plant height at harvest (cm)	Number of stems /hill	Number of tubers /plot	Weight of tubers /plot (kg)	Average of per tuber wt. (g)	Tuber yield (q/ha)	Marketable yield (%)
Control + *K.J.	81.5	6.6	71.3	2.3	218	16.4	75.1	255	83.1
Control + *K.C.	84.5	9.8	79.2	3.2	210	17.6	83.8	275	87.7
White polythene + K.J.	90.4	17.6	80.7	4.1	225	19.7	87.5	307	91.6
White polythene + K.C.	94.3	20.5	89.2	6.1	217	20.7	95.4	323	95.0
Black polythene + K.J.	95.5	22.1	93.3	6.5	226	21.0	93.8	328	95.0
Black polythene + K.C.	99.8	28.3	105.9	8.3	221	23.9	108.8	373	97.9
C D (0.05)	1.4	1.6	2.8	0.5	3.1	0.43	2.3	7.0	0.9

* K.J. = Kufri Jyoti
* K.C. = Kufri Chandramukhi

CHAPTER 7

Potato Production in Greenhouse

Plate 12: Early potato production in trench greenhouse

Effect of Planting Time on Yield of Potato in different Greenhouses

Potato is an important vegetable crop in the cold desert regions. But severe cold during winter season restrict its production in summer months (May-Sep). Defence Institute of High Altitude Research has done lot of research work on cold desert greenhouse technology in past one decade. The technology has been disseminated to the farmers of Ladakh region by way of demonstration and training. On the basis of shape, size and cladding material the Institute has designed/ developed eleven types of greenhouses. To maintain the availability of locally grown potato during June to August, an experiment was conducted to produce early (off season) potato in these greenhouses to select suitable greenhouse structure and planting date.

Nine treatment combinations (Table 7.2) consisted three type of greenhouses i.e. Trench greenhouse (TG), Fiber Reinforce Polyester greenhouse (FRP) and Double Wall Polyench (DWP) greenhouse, and three planting dates i.e. 31 January (T_1), 15 February (T_2) and 2 March (T_3) were tested for potato production during 2003. The seed tubers (about 40–50 g) of potato Cv. Kufri Chandramukhi were planted at 50 cm inter row and 15 cm intra row spacing. Other cultural practices were followed as per the normal potato crop grown in fields. The data on plant height, number of tubers per plant, weight of tubers per plant, average tuber weight and tuber yield (kg/m^2) were recorded at the time of harvesting. The trench greenhouse was managed as per described in subsequent experiment and the data on minimum and maximum temperature was recorded for all types of greenhouses during the cropping period (Table 7.1).

The data presented in Table 7.2 showed that the type of greenhouse and planting date have influence on all the parameters studied. Higher plant height was recorded in the

Table 7.1: Average minimum and maximum temperatures in different greenhouses during the cropping period

Months	Trench		Polyench		FRP		Open	
	Min.	Max.	Min.	Max.	Min.	Max.	Min.	Max.
January	-3.2	15.9	-1.2	19.9	-4.5	18.3	-13.8	2.1
February	-2.0	23.5	-0.8	23.1	-3.2	21.7	-11.1	5.2
March	2.3	33.1	3.6	30.6	-2.8	22.6	-6.4	10.4
April	6.6	36.3	8.1	32.3	3.2	32.7	-0.5	16.5
May	10.2	18.4	12.6	32.8	11.6	36.5	3.8	18.5
June	8.6	24.6	14.6	33.6	12.8	38.8	7.6	23.5

crop planted in FRP greenhouse on 02 March (95.6 cm) and was at par with the crop planted on 15 February and 31 January in the same greenhouse but was significantly higher than other type of greenhouses. Significantly more plant height under the FRP greenhouse could be due to the higher temperature recorded in FRP greenhouse than the other type of greenhouses (Table 7.1). Number of tubers were higher in the crop planted in trench on 02 March (8.6) and it was at par with crop planted in trench on 15 February (8.0) and significantly superior to other treatments. The more number of tubers per plant recorded in the planting on 02 March in trench greenhouse might be due to the trench greenhouse having more favorable temperature during tuberization stage (Table 7.1). Singh *et al.* (2000) reported 10–16^{0} C as the optimum temperature for better tuberization. Average per tuber weight was the highest with TG–T_2 (49.6 g) treatment and was statistically at par with TG–T_3 (46.9 g) treatment and significantly higher than other treatments. Higher per tuber weight observed in TG–T_2 may be due to the more translocation of carbohydrates to the tubers under more favorable temperature in trench greenhouse available for longer period (Table 7.1).

Table 7.2: Biological parameters as influenced by the types of greenhouse and planting date

Planting dates	Type of Greenhouse			
	Trench (TG)	FRP	DWP	Mean
	(1) Plant height (cm)			
31 Jan (T_1)	63.0	93.7	73.6	76.7
15 Feb (T_2)	63.7	95.0	76.3	78.3
2 Mar (T_3)	63.3	95.6	75.0	77.9
Mean	63.3	94.7	74.9	77.6
	(2) No. of tubers per plant			
31 Jan (T_1)	6.6	5.3	6.6	6.1
15 Feb (T_2)	8.0	6.0	7.6	7.2

contd...

table contd...

2 Mar (T_3)	8.6	6.7	7.3	7.6
Mean	7.7	6.0	7.1	6.9
	(3) Wt. per tuber (g)			
31 Jan (T_1)	41.8	35.3	34.1	37.0
15 Feb (T_2)	49.6	32.2	35.3	39.0
2 Mar (T_3)	46.9	32.0	39.4	9.4
Mean	46.1	33.1	36.2	38.4
	(4) Wt. of tubers per plant (kg)			
31 Jan (T_1)	0.27	0.18	0.23	0.22
15 Feb (T_2)	0.39	0.19	0.27	0.28
2 Mar (T_3)	0.41	0.21	0.29	0.30
Mean	0.35	0.19	0.26	0.26
CD (0.05)	(1)	(2)	(3)	(4)
	1.5	0.7	4.9	0.05

The higher weight of total tubers per plant recorded under the treatment TG–T_3 (0.41 kg) was at per with TG-T_2 and significantly superior over the other treatments. The superiority of the treatment TG–T_3 could be attributed to the more number of tubers per plant obtained.

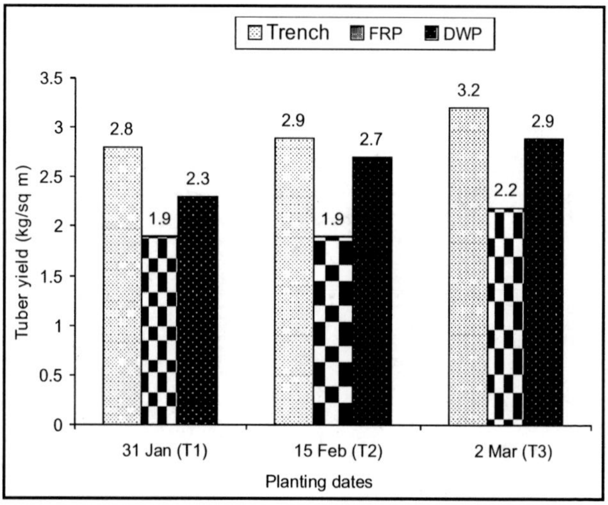

Fig. 7.1. Tuber yield (kg/ sq m) as affected by the type of greenhouse and planting dates

Maximum total yield was recorded with TG–T_3 (3.2 kg/ m^2) and was significantly higher than other treatments (Fig. 7.1). Higher total tuber yield (per m^2) under the treatment TG–T_3 was because of more number of tubers per plant and hence higher total tuber yield per plant. From the results it seems that tubers of Kufri Chandramukhi planted on 02 March in trench greenhouse (TG–T_3) gave maximum total tuber yield for early (off season) potato production in high altitude cold desert climatic conditions.

Effect of Harvesting Time (Crop Duration) on off Season Potato Production in Trench

The potato possesses a high production potential with adoption of suitable cultural practices on a scientific basis. Planting and harvesting time of potato crop are the two important factors which determine the tubers yield. Khurana and Bhutani (2003) reported that the total tuber yield increased with increase in crop duration. In earlier studies, last week of February to first week of March was observed to be optimum time of planting for early crop in trench. The present investigation was aimed at working out the optimum harvesting time to get higher yield of potato in trench greenhouse.

An experiment was carried out during 2000 and 2001 to find out optimum time of harvesting to obtain higher yield in trench with six harvesting dates (15 days interval) i.e. 6 June, 21 June, 6 July, 21 July, 5 August and 20 August of each year. 45 g seed tubers of Cv. Kufri Chandramukhi was planted in well prepared beds in trench (size 9 m length, 3 m width and 0.90 m deep) in the first week of March (5–6 March) in complete randomized block design with four replications. The tubers were planted with inter and intra row spacing of 50 and 20 cm, respectively. All cultural practices were followed as for ware potato crop in the region. Before planting, the seed tubers were taken out from underground store in second week of February and kept at a warmer place (in greenhouse) to facilitate emergence of small sprouts. The sprouted tubers were planted in the trench at 10–12 cm depth in furrows followed by 10–12 cm height ridges were made on the furrow. After planting of the tubers, the trench was covered by U.V. stabilized 200 micron thick translucent polythene sheet. To remove the excess heat from the trench during April–May the polythene was opened during day time and completely removed in second week of May. The crop was harvested at 90, 105, 120, 135, 150 and 165 DAP. Number of tubers per plant, average per tuber weight and total tuber yield (kg/m^2) were recorded at each harvest stage.

The pooled data of both the years (Table 7.3) showed that number of tubers per plant, average per tuber weight and tuber yield (kg/m^2) were significantly influenced at different harvesting days. Maximum number of tubers per plant (11.2) was recorded when the crop was harvested at 165 DAP and was at par to the crop harvested at 150 DAP (11.0) and both were significantly superior than other treatments. The maximum number of the tubers at 150 and 160 DAP could be due to increase in the tuber size in due course of time in the favorable climatic conditions of the region, which was not observed in earlier harvests. Likewise, increase in per tuber weight was recorded with increase in time for harvesting. Higher tuber weight was recorded at 165 DAP (59.7 g) but no significant increase obtained in tuber weight was after 135 days of planting.

Table 7.3: Number of tubers per plant, average tuber weight and tuber yield of potato crop as affected by harvesting time (pooled over 2000 -01)

Treatment	No. of tubers/plant	Av. wt. per tuber (g)	Tuber yield (kg/m^2)
Harvesting dates			
6th June (90 DAP)	6.0	40.2	2.4
21st June (105 DAP)	7.4	42.1	3.1
6th July (120 DAP)	8.4	46.4	3.9
21st July (135 DAP)	9.5	56.5	5.3
5th August (150 DAP)	11.0	58.3	6.1
20th August (165 DAP)	11.1	59.7	6.2
CD (0.05)	0.4	1.3	0.13

Higher tuber yield 6.2 kg/m^2 observed with delayed harvesting (165 DAP) of the crop was statically at par with crop harvested at 150 DAP. The higher tuber yield recorded may be due to increase in tuber number and tuber weight with crop stand was for a longer time. Singh *et al.* (2003) also reported increased tuber yield with longer crop duration. The study revealed that although, tubers for table purpose can be taken from the crop in first week of June (90 DAP). But significantly higher tuber yield can be obtained by potato planted in first week of March in trench and harvested after 150 days of planting.

Evaluation of Different Varieties for off Season Potato Production in Trench Greenhouse

To maintain round the year availability of locally produced potato, the trial was conducted on early potato production in various type of greenhouses at different planting dates. After selecting the suitable greenhouse (trench) and planting date (last week of February to first week of March), there was a need to evaluate more varieties to get maximum yield in trench greenhouse. Therefore, an investigation was carried out to find out the suitable potato variety for early (off season) potato production in trench using newly released varieties obtained from CPRI, Shimla.

The seed tubers (about 40-50 g) of five potato Cvs. Kufri Giriraj, Kufri Jyoti, Kufri Chandramukhi, Kufri Jawahar and Kufri Sultan, and same size seedling tubers of hybrid TPS strain I/13 were planted at 50 x 15 cm spacing on 5 March, 2004 in trench greenhouse with four replications. All other cultural practices were followed as for ware crop in the region. The crop was harvested on 15 June, 2004 (100 DAP).

The results (Table 7.4) indicated that all the parameters studied were significantly influenced by the treatments. Maximum plant height was recorded in Kufri Sultan (78.3 cm) that was at par with Kufri Giriraj (74.3) and both were significantly higher than others. More plant height in Kufri Sultan may be due to the its genetic makeup for the growth of the plant in favorable climatic condition. Significantly higher number of tubers per plant were recorded in TPS strain I/13 (7.7). The variety Kufri Giriraj, Kufri Jyoti, Kufri Jawahar and Kufri Sultan were statistically at par in term of numbers of tubers per plant. Minimum number of tubers per plant were recorded in Kufri Chandramukhi (5.5). Maximum total tuber yield per plant was recorded with Kufri Chandramukhi (0.47 kg) followed by Kufri Giriraj (0.43 kg) and were significantly superior to other cultivars. Likewise, Kufri Jyoti (0.38 kg), Kufri Jawahar (0.36 kg), TPS-I/13 (0.37 kg) and Kufri Sultan (0.36 kg) yielded at par.

Table 7.4: Growth and yield performance of different potato cultivars and TPS seedlings in trench greenhouse

Treatment	Plant height (cm)	No. of tubers /plant	Wt. per tuber (g)	Yield /plant (kg)	Tuber yield (kg/m²)
Variety					
Kufri Giriraj	74.3	6.4	68.0	0.43	4.27
Kufri Jyoti	61.6	6.6	57.5	0.38	3.90
Kufri Chandramukhi	59.6	5.5	77.2	0.47	4.90
Kufri Jawahar	60.0	6.7	52.9	0.36	3.63
TPS-I / 13	68.0	7.7	37.0	0.37	3.70
Kufri Sultan	78.3	6.3	56.8	0.36	3.60
CD (0.05)	2.6	0.4	1.4	0.06	0.19

Average tuber weight was also significantly influenced in different cultivars (Table 7.4). More tuber weight was recorded with Kufri Chandramukhi (77.2 g) followed by Kufri Giriraj (68.0 g). Variety Kufri Chandramukhi was statistically superior than others which may be due to the effecient carbohydrates accumulation in tubers resulted by increased size of the tubers. Maximum total tuber yield per m^2 recorded in Kufri Chandramukhi (4.90 kg) was statistically at par with Kufri Giriraj (4.27 kg) but significantly higher than Kufri Jyoti (3.90 kg), Kufri Jawahar (3.63 kg), TPS-I/13 (3.70 kg) and Kufri Sultan (3.60 kg). The results showed that variety Kufri Chandramukhi and Kufri Giriraj were found suitable for higher tuber yield per unit area in trench greenhouse.

❑❑❑

CHAPTER **8**

Storage Studies

India is a leading potato producer country and most of the potatoes are produced in Indo-Gangetic plains and harvested in the month of February-March, coinciding with rising ambient temperature. Rising temperatures result in sprouting, weight loss and rotting of tubers. Hence, the potatoes are stored in refrigerated cold stores at 2-4^0 C temperature during summer (Kumar *et al.* 1999; Bhutani and Khurana, 2005). This is required heavy expenditure to create the facility cold storage and continuous supply of electricity.

Unlike plains, potato storage facility (controlled environment) is not available in hilly regions like Ladakh, although more than 80% produce harvested from the potato crop during September–October is required to be stored to maintain regular supply of potatoes in the region. The potato crop is harvested in September-October followed by a period of severe dry cold weather till April and again relatively hot weather during May-August. Therefore, the storage facility is required at elevated temperatures during November-March/April (as temperature goes upto minus 30° C) and at lower temperatures during summer season (May-August). Due to shortage of energy source, the region has peculiar problem of storage and continuous supply of locally produced potatoes. Although, the region has potential to increase the potato production with adoption of modern cultural practices developed and recommended for the specific climatic conditions of the region. The problem related to storage was identified as the major constraint in the way of achieving further increase in potato production and its round the year availability for table purposes and also storing as a seed for the next planting season. To overcome these problems geothermal energy was used for storage studies and several research trials were conducted on storage of potato in underground store. The study revealed that potatoes can be stored successfully from November-April without much damage in zero-energy based underground storage facility. In this technique, geothermal energy and energy released by the tubers in respiration process increases the temperature of store and saves the tubers during freezing winter season.

A sprout suppressant chlorpropham or CIPC (isopropyl 3-chlorocarbanilate) was registered in India in 1998. It is commercialy used for potato storage at 10-12^0 C temperature in controlled environmental conditions of cold store. However, some alterations are required in the cold store for adoption of this technology, which involves heavy investment. In the peculiar problem of high altitudes, the experiments were conducted on long term potato storage in specifically designed underground store with CIPC treatment. As a outcome of these experiments, potatoes could be stored successfully from Nov-Aug. Complete exhibition of sprouting was observed is the stored potatoes. The details of these trials conducted under the storage studies programme are described in the subsequent chapters.

❑❑❑

8.1
Storage Using Geothermal Energy

(1) Traditional pit

(2) Modified pit

(3) Underground store

Plate 13: Different type structures for potato storage

Storage of Potatoes from TPS progenies & Varieties in different Storage Structures using Geothermal Energy during freezing Winers

In traditional method, potatoes are stored in deep pits at the suitable place available with farmers in open field conditions. They make 1.20-2.10 meter deep and 0.90-1.5 meter diameter pit depending upon the quantity of potatoes to be stored. These pits are filled up with potato tubers leaving 30-45 cm space from the surface. The pit is finally covered with 30-45 cm dry soil. The potato crop is generally harvested in September in the region. The potatoes are filled up in the pits from mid October to mid November depending upon the altitude and are stored up to March. But in this type of storage system, heavy losses occur due to sprouting, shrinkage and rotting of the tubers. It is general opinion that the losses occur due to low temperature during severe winter months (December-January) and high temperature in February-March inside the pit. To find out the actual reason of spoilage during storage, two structures, one modified pit and another underground room was constructed. The modified pit measured 2.40 m deep, 1.20 m long and 1.20 m wide was constructed with the help of unbaked mud bricks. The underground room was constructed measuring 2.40 m deep, 1.80 m long and 1.50 m wide with masonry stone walls. The roof of the room was made with the help of wooden poles, sticks, mud and dry soil. There was provision of one door (1.20 x 0.90 m) and one small ventilator at back wall (opposit of the door) of the room. The study was carried out by storing potato tubers harvested from different TPS progenies and variety in the traditional pit, modified pit and underground store.

The investigation was carried out by storing the tubers of different TPS progenies i.e. I/13, I/67, II/13, II/67, 7/67, II/TPS–FRL-1, TPS–FRL–2/67 and a variety Kufri Chandramukhi in different type of structures like traditional pit, modified pit and underground store during winter months (November-March). 50 kg tubers of each progeny and variety in four replications were stored in each structure. All the three structures were filled up in the first week of November and the tubers were kept in nylon netted bags. The pits were filled up to 1.80 meter height with tubers and than with 45 cm thick dry soil in the traditional pit and the modified pit. The bags were stacked on 10 cm diameter wooden poles at the bottom of the under ground room. The thermometer probes were kept between the bags at 2.4 m, 1.50 m and 0.60 m depth in all the three structures. The probes were connected with a battery operated multi-channel digital temperature recording machine. The temperature of all the probes was recorded three times every day at 9 PM, 5 AM and 1 PM in traditional and modified pit and average monthly temperature was calculated (Table 8.1.1). A traditional pit was made for control treatment and probes were kept at the same depth under the soil as per treatments without potato tubers and temperature was recorded in a similar manner. The potato tubers were taken out on 15th March from traditional pit and modified pit and also from the underground room. Replication wise percentage of weight loss, rotted tubers and sprouted tubers were recorded and data was statistically analysed by using single factor randomized block design. Daily maximum and minimum temperature and relative humidity were recorded in underground room with the help of digital thermo-hygrometer and was calculated on the basis of monthly average (Table 8.1.2).

During the storage period temperature varied slightly when it was recorded at 9 PM, 5 AM and 1 PM in the traditional pit and modified pit including traditional pit without potatoes (Table 8.1.1). Maximum temperature of 16.6^0 C was recorded at 0.60 m depth in modified pit at 9 PM during the month of December. The temperature in traditional pit was not much different at different times period of the day and slightly higher temperature (15.1^0 C) was recorded at 1.50 m depth at 9 PM in the month of February. During December to March the temperature difference was not much in both modified and the traditional pits. There was much difference in the temperature recorded at similar depth in control (without potato) and in traditional pits. There were large variations in temperature, observed at different depths in the control pit. Highest average temperature 8.7^0 C was recorded at 2.40 m depth in the coldest month of January and lowest -1.5^0 C at 0.60 m depth, however, temperature at 1.20 m in control pit was recorded as 4.0^o C. Much fluctuation in temperature specially at 0.60 m depth was recorded with fluctuation in out side (open) temperature and it was lowest during January and the highest in March.

Minor fluctuation in temperature was recorded in different months and during different times of the day in modified and traditional pits. Both type of pits had higher temperature than control, which may be due to the generation of heat through respiration of the tubers and insulation properties of the soil. High temperature in modified underground pit might be due to the heat generated in respiration, which was accumulated within the pit as the pit was made of mud bricks, however, heat generated in traditional underground pit might have been dispersed/transmitted to near by soil as the traditional pit had comparatively loose soil of vertical surface. During the period, minimum (-14.9^0 C) average open temperature was recorded in January at 5AM and maximum (14.3^0 C) in March at 1 PM. In both the type of pits much higher temperature were recorded as compared to the minimum temperature in open condition. It might be due the heat generation from the tubers and absorption and emission of heat tendency of the soil. The temperature increased with increasing depth in control pit, probably due to the reduction in penetration of low temperature with the increasing depth.

Table 8.1.1. Temperature regime during underground storage of potato in different types of pits

Structure	Depth of probe (m)	Dec 96				Jan 97				Feb 97				Mar 97			
		9PM	5AM	1PM	Av.	9PM	5AM	1PM	Av.	9PM	5AM	1PM	Av.	9PM	5AM	1PM	Av.
A.Modified	i) 2.4	16.1	15.5	15.8	15.8	14.9	14.7	16.0	15.2	13.1	13.1	13.2	13.1	13.5	13.1	16.3	14.3
underground pit	ii) 1.5	16.0	15.6	15.7	15.7	14.4	14.1	14.7	14.4	13.1	12.9	13.2	13.1	13.5	14.4	14.2	14.0
	iii) 0.60	16.6	16.2	16.3	16.3	15.0	14.8	15.4	15.1	13.6	13.6	13.8	13.7	14.3	14.1	14.7	14.4
B.Traditional	i) 2.4	11.9	11.5	11.7	11.7	10.8	10.6	11.0	10.8	9.7	9.4	9.4	9.5	9.5	9.3	9.7	9.5
underground pit	ii) 1.5	13.1	12.6	12.9	12.8	13.7	13.6	14.0	13.8	15.1	13.7	14.1	14.3	14.0	13.6	14.3	13.9
with potato	iii) 0.6	11.3	11.0	11.4	11.2	11.0	10.7	11.2	10.9	11.4	11.2	11.3	11.3	12.0	11.6	12.5	12.0
C. Traditional under	i) 2.4	9.0	8.8	9.2	9.0	8.7	8.6	8.8	8.7	7.8	7.3	7.5	7.5	7.3	8.0	7.6	7.6
ground pit (control	ii) 1.5	4.4	4.2	4.6	4.4	4.0	3.9	4.1	4.0	3.8	3.9	4.1	3.9	5.4	6.6	5.5	5.8
with out potato)	iii) 0.60	-1.0	-0.9	-0.8	-0.9	-1.5	-1.4	1.3	-1.4	0.2	0.2	0.4	0.3	3.7	3.8	4.0	3.8
D.Traditional pit of	i) 0.60	4.0	3.6	5.2	4.3	3.7	3.6	4.6	3.9	4.9	4.8	5.9	5.2	7.2	6.8	8.5	7.5
4′ depth	ii) 1.2	9.4	8.9	11.1	9.8	8.3	7.8	9.5	8.5	10.1	10.0	12.5	10.9	11.5	12.1	12.9	12.2
E- Open (Outside)		-4.5	-11.4	5.2	-10.7	-9.4	-14.9	8.0	-10.8	-5.5	-13.6	4.8	-4.8	4.8	-7.6	14.3	3.8

(All values: Average Temperature °C)

When total depth of the traditional pit was kept 1.20 meter and potato tubers were stored in it, the change in temperature behavior was noticed and it was similar to refrigeration condition. At 1.20 m depth, temperature was recorded higher than 0.60 m but the temperature was lower at the same depth than modified and traditional pits of 2.40 m deep.

The average maximum, minimum temperature and relative humidity data (Table 8.1.2) in under ground store fluctuated during different winter months. Temperature decreased with decreasing outside temperature up to January and again increased with increasing open temperature from February to April. It was recorded to be 18.9^0 C as maximum and 10.7^0 C as minimum in October and both maximum and minimum temperatures were the lowest in the month of January. 1.5^0 C (minimum) temperature was recorded inside the store when outside temperature was -14.9^0 C in the month of January. The lowest (77.7%) relative humidity was recorded in the month of October and the highest (97.9%) in January in the underground store. There is also a little scope to regulate the temperature and humidity by adjusting in the time for opening/closing of ventilator and door of this underground store.

Table 8.1.2: Average minimum and maximum temperature (^{0}C) and RH in underground store during potato storage in winter months

Months	Minimum		Maximum		RH (%)	
	Store	Open	Store	Open	Store	Open
October	10.7	5.4	18.9	13.3	77.7	50.3
November	9.4	-6.9	14.3	8.1	92.1	55.4
December	3.6	-10.6	6.4	2.6	97.2	64.6
January	1.5	-13.5	2.5	2.0	97.9	64.9
February	2.2	-9.3	3.8	5.8	96.2	60.6
March	4.1	-6.4	7.3	6.6	92.3	58.4
April	6.8	3.5	11.6	15.5	89.3	49.3

Data from Table 8.1.3 indicated minimum weight loss of tubers in underground store and the highest in modified pit. Storage in traditional pit showed lesser weight loss than the modified pit, but major difference was observed in the underground store. Similar trend was observed in term of rotting and sprouting of tubers. In term of weight loss, rotting and sprouting, underground room storage was observed to be superior and it may be due to the more favorable temperature and humidity for the potato tubers stored in it during winter months.

There was significant difference in TPS progenies/variety and storage structure on weight loss, rotting and sprouting of tubers (Table 8.1.3). Maximum weight loss was recorded in II/13 in traditional pit. Significantly minimum weight loss was recorded in TPS-FRL-2/67 than all other TPS progenies and variety Kufri Chandramukhi (control). The highest rotting was recorded in Kufri Chandramukhi followed by II/TPS-FRL-1 in traditional pit. TPS progeny II/TPS-FRL-1 recorded the highest sprouted tuber percentage followed by Kufri Chandramukhi. TPS-FRL-2/67 recorded minimum weight loss, sprouted tubers and rottage percentage in traditional pit storage and was significantly superior over other TPS progenies/variety. Perhaps, increased temperature in modified pit than the traditional pit, resulted in increased weight loss, rotting and sprouting in all the TPS progenies and variety. Significantly minimum weight loss, rottage and sprouting were recorded in TPS-FRL-2/67 in modified pit, however, maximum weight loss percentage was observed in II/TPS-FRL-1 and rotting and sprouting in Kufri Chandramukhi (control) in modified pit. Kumar *et al.* (2001) and Singh *et al.* (2001) have also reported variation in shelf life of the different potato varieties when they were stored at ambient temperatures. When the tubers of TPS progenies and variety were stored in underground store, less percentage of weight loss, rotting and sprouting were recorded than the traditional and modified pits. The difference in weight loss percentage between the TPS progenies / variety was almost identical in the underground store. More percentage of rotted and sprouted tubers were recorded in Kufri Chandramukhi than all the TPS progenies in underground store. Significantly minimum rotting and sprouting in TPS-FRL-2/67 was recorded in underground store during the period. The maximum weight loss, rotting and sprouting observed in modified and traditional pits compared to underground store, may be due to the fact that these structures had higher temperature than underground store during the storage period (Table 8.1.1). Mehta and Kaul (1988) and Verma *et al.* (1974) also reported higher weight loss, rotting and sprouting at higher temperatures. The underground store maintained more favourable temperature and humidity (Table 8.1.2) due to the proper harnessing of geothermal energy and energy released by the tubers, which was maintained manually due to its design having the provision of ventilator and door. One MT potato stored in 160 cubic ft. (8′x5′x4′) in underground store has been proved suitable to maintain favourable temperature during winter months.

From the above results, it can be concluded that potato tubers can be stored successfully with acceptable weight loss, rotting and sprouting during freezing winter months (October-March-April) in high altitude cold desert conditions. Among the TPS progenies/variety, II/67 was found the best for storage. Underground store was found to be a suitable structure to maintain favorable conditions for potato storage. The structure has also got certain advantages than the modified and the traditional pit like (1) provision for proper harnessing of geothermal energy, (2) the temperature and humidity can be modified by manually opening of the ventilator and door, (3) potato tubers can be taken out from the room store at any time, (4) potato tubers stored in this type of store remains neat and clean and (5) potatoes can be stored up to the month of April without much damage by sprouting, weight loss and rotting in this specially designed underground store.

Table 8.1.3. Percentage of weight loss, rotted and sprouted tubers during storage in different storage structures in various potato cultures during winter

Potato cultures	Traditional pit			Modified pit			Underground store		
	Wt. loss (%)	Rotted tubers (%)	Sprouted tubers (%)	Wt. loss (%)	Rotted tubers (%)	Sprouted tubers (%)	Wt. loss (%)	Rotted tubers (%)	Sprouted tubers (%)
I / 13	9.1	8.6	26.6	10.2	9.3	30.5	5.3	4.6	7.4
I / 67	8.9	8.1	24.9	10.9	8.4	28.7	4.8	4.2	6.2
II / 13	10.8	8.1	27.3	11.2	8.5	32.1	6.6	5.4	7.8
II / 67	7.8	6.0	24.3	9.2	7.5	26.0	4.3	4.3	6.9
7 / 67	6.3	6.7	23.5	8.8	7.7	23.8	5.4	2.8	5.5
II / TPS - FRL - 1	8.9	10.0	32.6	13.9	10.6	45.1	6.5	5.5	8.3
TPS - FRL - 2 / 67	3.5	3.4	18.2	5.5	3.7	25.2	5.1	2.3	3.6
Kufri Chandramukhi	7.5	12.9	30.9	9.1	13.6	47.3	6.2	8.2	9.4
CD (0.05)	0.34	0.25	0.91	0.39	0.57	0.69	0.51	0.40	0.31

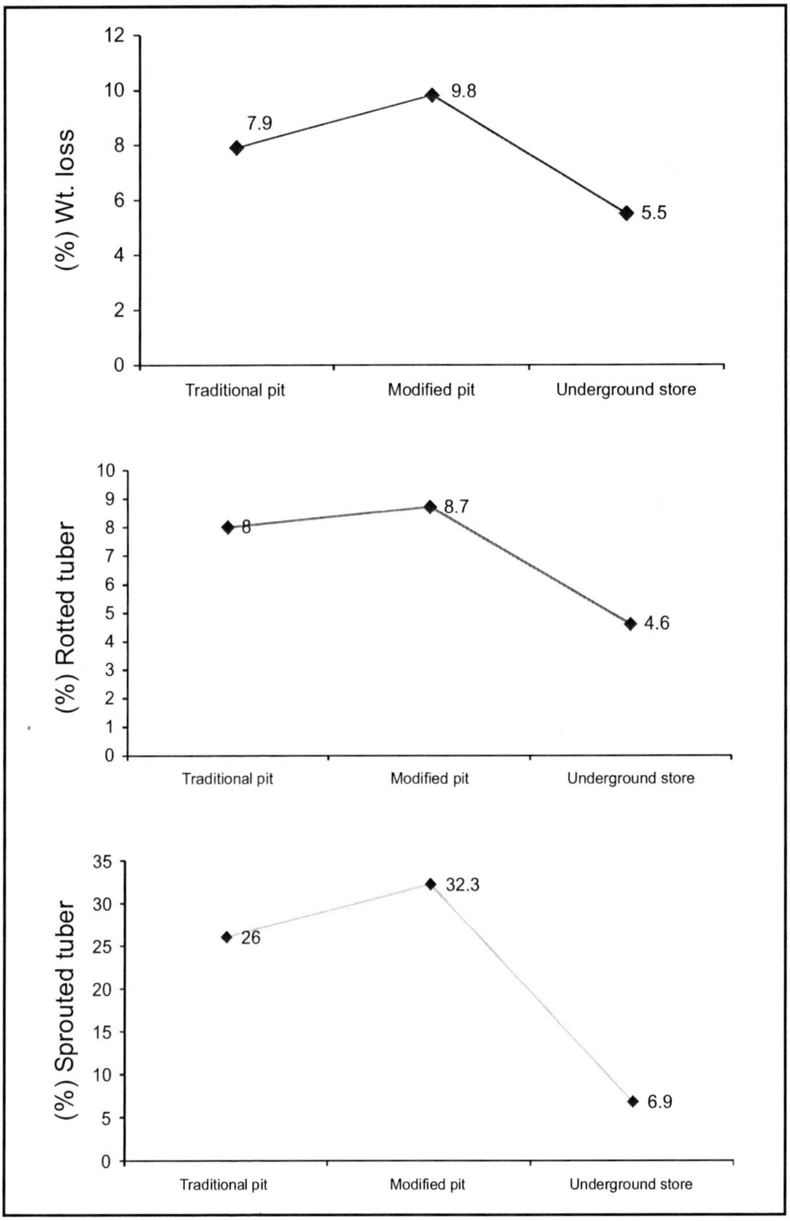

Fig.8.1. Average % of wt. loss, rotted and sprouted potato tubers during storage in different structures.

8.2
Storage with CIPC

Different packing of potato tubers

Potato storage with different packing in underground store

Layout of bags in store for trial

Handy fogging machine

10 Months stored potato tubers control and CIPC treated

Plate 14

Studies on Zero - energy based Long-Term Potato Storage with CIPC Treatment

Potato is available in high altitude from September-October from the freshly harvested crop. To store these potatoes from November to March, an underground storage technique has been standardized as described in previous chapter number 8.1. Heavy sprouting, weight loss and shriveling were the main problems in further storage from May-August. Therefore, to meet the requirements from May-August, the people of the region depend on potato procured from other parts of the country. However, the region has potential to produce potatoes to fulfill round the year local requirement. But after the month of April, apparently due to physiological age of the potatoes, increase in temperature and decrease in humidity, the potatoes sprout due to which shriveling takes place making them unsuitable for consumption. Therefore, sprouting and low humidity were major constraints for reducing the storage life and quality of the produce. Use of sprout suppressants and provision of adequate humidity may help in controlling the sprouting and shriveling to increase the storage life of potatoes. Chloropropham or CIPC (isopropyl 3-chlorocarbanilate) is the world-wide most widely used sprout suppressant for storage of ware potatoes. CIPC acts on the growing tips of the sprouts and prevents cell multiplication therefore, stops growth of the new sprouting shoots and roots which is the major problem for long term potato storage in high altitudes. The present study was, therefore, undertaken to evaluate the efficiency of CIPC for the suppression of sprouting in potatoes stored in geothermal energy based specifically designed under ground store at 3450 above mean see level.

The experiment was conducted on potato variety Kufri Chandramukhi stored with CIPC treatment. Potatoes were filled in different types of packings such as Netted plastic baskets, Nylon netted bags, thin netted Gunny bags and thick Gunny bags. 25 kgs of potatoes were kept in plastic buskets and 50 kgs tubers were filled in other type of packings in three replications. All the baskets and bags were kept in an underground store in second week of October. The fog of CIPC chemical (Oorja of United Phosphorus Ltd, Mumbai containing 50% a.i.) was applied by a fogging machine at the rate of 60 ml per MT of potatoes plus 30% (likage). The door and ventilators of the underground store were sealed for next 48 hours of fog application. The first fogging was done during second week of October 2006 and second fog application during last week of March 2007. The potato having single spray treatment were kept in same type of another store with control before application of second treatment. The door of the store was opened for about half an hour for recording weekly temperature and humidity data (Table 8.2.1). The observations were recorded on sprouting, weight loss and shriveling of the tubers in the second week of March 2007 and there was no sprouting and rotting observed in the treatments including control. Second observation was recorded in the last week of August 2007. The residues of CIPC and reducing sugars were analyzed in treated, untreated and freshly harvested potatoes.

Table 8.2.1: Average temperature and relative humidity in under ground store during different months

Month	Temperature (°C)				RH (%)	
	Maximum		Minimum			
	Store	Open	Store	Open	Store	Open
Oct	8.2	14.9	6.7	-1.8	81.2	50.4
Nov	7.7	9.7	5.6	-8.4	82.6	55.4
Dec	6.8	5.0	4.4	-12.7	79.7	64.6
Jan	7.7	-2.0	0.8	-13.6	84.6	64.9
Feb	7.1	4.3	1.3	-10.5	88.2	60.6
Mar	7.7	8.1	2.6	-6.1	97.0	58.4
April	10.5	14.8	8.4	-0.4	96.8	49.3
May	14.0	18.4	11.5	3.8	91.0	52.8
June	18.6	23.5	14.2	7.9	88.5	48.9
July	20.5	29.3	18.7	12.0	82.2	45.5
Aug	22.6	28.1	19.5	10.9	85.2	54.8

Data from the table 8.2.1 revealed that the overall minimum temperature 8.5^0 C was recorded inside the specially designed store during the storage period as compared to - 1.7^0 C ambient temperature. Therefore, the design of the store helps to increase average temperature 10.2^0 C during the peak winter period, which save the potato tubers from spoilage due to freezing at low temperature (below 2^0 C). During peak summer months maximum temperature 11.9^0 C was recorded inside the store and it was 3^0 C less from ambient. Lowering the temperature is helpful to keep the tubers at right physiological stage for longer period. Humidity is influenced mainly to evapotranspiration rate, weight loss and shriveling and required adequate (about 90.0%) humidity during storage. 87% humidity was achieved through provisons of water channels and watering inside the underground storage structure, whenever, ambient humidity was recorded 55%. Therefore, this design of the store helps to increases humidity by 32% to ambient. The comparative data of temperature and humidity of the store and ambient recorded during the experimentation period is represented in Table 8.2.1.a.

Table 8.2.1.a. The overall average maximum and minimum temperature and humidity in the store and ambient during the storage period.

S. No.	Parameters	Store	Ambient
1.	Maximum temperature (^{0}C)	11.9	15.0
2.	Minimum temperature (^{0}C)	8.5	-1.7
3.	Humidity (%)	87.0	55.0

CIPC suppressed sprouting in potatoes up to 300 days (10 months) after first fog application (Table 8.2.2). Sprouting was started in first week of April (150 days of storage) in untreated tubers (control), and reached to cent percent level after 180 days of storage. However, the sprouting was observed to be only 2.7 % in single and 1.9 % in double fogging treatment in thin gunny bags, whereas it was higher 16 % in single and 12.3 % in double treatment when the tubers were packed in thick gunny bags however, there were no sprouting in plastic basket and Nylon netted bag packing in both the fogging application treatments. There was not found much difference in sprouting with single and double fogging treatments. The control of sprouting was better in plastic baskets, nylon netted bags and thin netted gunny bags compared to thick gunny bags, probably due to better reaching of fog on the tuber surface and thick gunny bags checks the fog to enter in it.

The data on weight loss taken during last week of August (300 days after first fog application) revealed that minimum weight loss was recorded when treated potatoes were kept in thick gunny bags than thin netted gunny bags, nylon netted bags and plastic baskets. Higher weight loss in untreated (control) treatment was attributed to profuse sprouting in the tubers (Table 8.2.2). The minimum weight loss observed in thick gunny bags was could be due to the insulation and checking of the moisture loss from the surface of the tubers than in other packings, as the humidity fluctuates naturally with the weather condition in the store during the period.

Table 8.2.2: Effect of CIPC on sprouting, weight loss and shriveling percentage during potato storage in underground store

Treatment	Single spray			Double spray		
	Sprouting (%)	weight loss (%)	Shriveling (%)	Sprouting (%)	weight loss (%)	Shriveling (%)
1. Untreated (control)	100	38.6	100	100	38.6	100
2. Plastic Baskets + CIPC	00	13.2	40.2	00	14.0	67.8
3. Nylon Netted bags + CIPC	00	9.2	6.2	00	8.8	6.3
4. Thin netted Gunny bags +CIPC	2.7	6.4	5.6	1.9	6.3	5.2
5. Thick Gunny bags + CIPC	16.0	5.0	3.2	12.3	4.2	3.8

The data showed 100% shriveling in untreated potatoes. However, it was minimum in thick gunny bag packing than in thin netted gunny bag, nylon netted bag and plastic baskets. The minimum shriveling was recorded in gunny bag packing, which may be attributed to better insulation between tubers and environment in these bags. The shriveling caused considerable losses in control tubers and they were become unfit for marketing/consumption, whereas more than 95% of CIPC treated potatoes were fit for marketing/consumption and had good appearance when they were kept in Nylon netted bags and thin gunny bags.

The residue concentration of CIPC was analyzed with the help of Punjab Horticulture Technology Centre, Campus Punjab Agriculture University, Ludhiana and it was found to be below detection level (Minimum Detective Level 0.04 mg/kg) in both the peeled and unpeeled samples collected from the gunny bags. In the samples collected from the plastic baskets, the CIPC residue was observed to be 1.28 mg/kg in peeled potatoe's and 118 mg/kg in unpeeled potatoes. The higher level of CIPC residues found in unpeeled potatoes collected from plastic baskets might have been due to the fact that the chemical was directly in contract with the skin of the potato tubers. But after peeling the tubers, residue was found to be far below the minimum residue level recommended by environmental protection agency (EPA) as 30 mg/kg in potatoes (EPA, 1996). This indicates that the treated potatoes used after peeling or without peeling are safe for human consumption when they stored in gunny bags. When the chemical was applied directly on the tubers they were safe for consumption after peeling. Singh and Ezekiel (2005) also reported very low residue level in stored potatoes with CIPC treatment.

Reducing sugar level was observed to be 0.46 % in the stored tuber samples. Reducing sugar content upto 0.25% is considered acceptable for producing chips. However, for French fries and dehydrated products, reducing sugar content upto 0.5% is considered acceptable (Singh *et al.*, 2002). Therefore, the stored potatoes were suitable for processing to make dehydrated products and French fries. But to make quality chips, the reducing sugar level was observed to be higher. The low sugar level may be maintained by harvesting the crop at proper time and avoiding very low temperature during storage.

The study revealed that locally produced potatoes can be stored from October to August without much damage in underground store with CIPC treatment. It will helpful to make the region self sufficient in term of round the year availability of locally produced potatoes.

❑❑❑

CHAPTER **9**

Package of Practices

(1) Leaf

(2) Flower

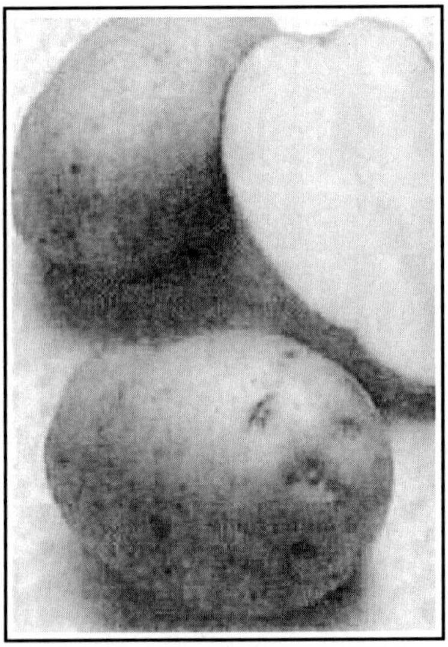

(3) Tuber

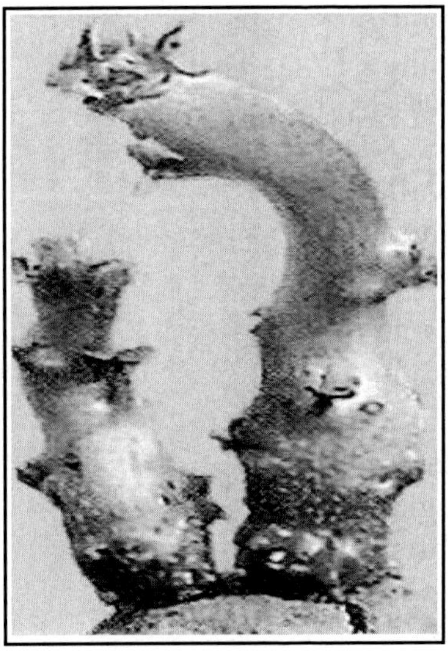

(4) Sprouts

Plate 15: Feature potato variety of Kufri Chandramukhi

Agro-Practices Based on the Findings of the Experiments

The climatic and geographical conditions are adverse in high altitude cold arid region. The temperature is unstable, unpredictable and fluctuation in seasonal and diurnal temperature is very high (Table 11). Range of minimum and maximum atmospheric humidity is also very high. It is also depected from the climatic data given in Table 11 that the sudden drop in temperature during early stage (April/May) and later stage (September/October) of crop growth is a common phenomenon. Wide variation in climatic conditions make it difficult to standardize package of practices for potato crop in the region. The soils formed by the weathering of different type of rocks (Plate 14.1) in the different locations vary in physico-chemical composition. However, in high altitude regions, large number of villages situated at > 3600 meter altitude have short cropping season (Table 1.2). The people of the region depend only on barley at higher altitude and on wheat and barley at lower altitude is staple food. Potato being a short duration staple food crop, can play an important role for food and nutritional security and eradicating the poverty and hunger in this remote region. However, due to non availability of data base, quality seed of suitable variety, standard package of practices and storage techniques, the potato crop has not been able to pickup movement in the region.

To boost its production, availability of locally produced seed and make the region self sufficient in potato, the research was carried out on different aspects of the potato cultivation. Research findings over one decade may help in making the region self sufficient, and supply of surplus quantity to other parts of the country.

Soil

Samples were collected from 96 villages of Leh and 50 villages of Kargil district. Depending upon the land holding, 3–6 samples were collected from each village. The samples were analyzed at the Soil Science Division, IARI, New Delhi and Dr. Y. S. Parmar University of Horticulture and Forestry, Nauni, Solan (HP). The data revealed that the type of soil in the region depends upon weathered rocks in the particular location and soils are sandy, sandy loam, clay loam, silt clay or clay. Majority of the soils are sandy to sandy loam in texture and low in organic matter with poor water holding capacity. The pH ranged from 5.65 to 10.12 in Leh and pH of Kargil varied from 6.57- 9.47 (Table 9.1). More than 90% samples from both the districts had pH in the range of 7–9. Therefore, majority of the soils in high altitude cold desert regions are alkaline in nature. Electric conductivity ranged from 0.05–1.56 mmos/cm in district Leh and 0.08–1.55 in Kargil. More than 90% soils were low in available phosphorus but high in potassium. Availability of the potassium depends upon the other parameters of the soil. It was observed that additional dose of potassium had a positive impact not only on potato but also on yield of other vegetable crops. More than 30% samples were low in micro-elements therefore may led to deficiency symptoms in the crop like wheat, barley and vegetables including potato (Singh and Singh, 2003). Limited patches of clay soil are available in the region, which were found not suitable to grow potato crop. Sandy loam soils of the region with proper dose of organic manure are the best for higher yield and better quality of the produce.

Table 9.1: Range of chemical properties of soils of different villages in Leh and Kargil district

Parameters	Leh		Kargil	
	Minimum	Maximum	Minimum	Maximum
pH	5.65	10.12	6.57	9.47
EC (m mos/cm)	0.05	1.56	0.08	1.55
Organic matter (%)	0.01	1.41	0.21	3.2
Phosphorus (kg/ha)	4.0	25.66	2.80	116.1
Potash (kg/ha)	12.32	496.15	115.0	861.3
Zinc (ppm)	0.06	6.0	0.14	5.17
Iron (ppm)	0.54	33.79	0.20	38.1
Copper (ppm)	0.43	3.52	0.48	3.41
Manganese (ppm)	0.2	32.17	0.40	4.74

Thus, potato can be grown successfully in sandy and sandy loam soils of the region with comparatively higher dose of organic manures and fertilizers and requisite number of irrigations.

Land preparation

Although, not much work has been done on the number and depth of ploughing in high hills in term of working out expenditure on ploughing of the field and tillage operations in potato crop. In high altitude cold desert conditions, one ploughing is required in September after harvesting the summer crop. One turning to the soil just before onset of severe winter (November/December) is beneficial to reduce the incidence of soil born diseases and insect-pests and to control the weeds to some extent because during December-January the temperature reaches down as minus 20^0 C in the regions. Therefore, the most of the pathogens and different insect-pests can not survive at this low temperature. After winter, in April/May depending upon the altitude, one deep (about 15 cm) ploughing followed by planking was sufficient to get the optimum potato yield in the region.

Variety

Selection of suitable variety is the most important factor to maximize the potato yield in a particular climatic condition. Results showed that Cv. Kufri Chandramukhi recorded maximum tuber yield and the data from earlier studies also revealed Kufri Chandramukhi the most suitable cultivar followed by Sindhu Red and Kufri Giriraj. Kufri Giriraj also gave higher marketable tuber yield and was resistant to common scab. Kufri Jyoti has early bulking than other varieties but was observed to be more susceptible to common scab followed by Kufri Chandramukhi. The Kufri Jyoti was also more suceptable to potato cutworm as compared to other varieties. The Kufri Chandramukhi is recommended for cultivation in cold desert regions of Lahaul and Spiti (Anonymous) and Afghanistan (Kadian *et al.*, 2003). Therefore, it can be concluded that Kufri Chandramukhi, Kufri Giriraj and Sindhu Red are the suitable varieties for high altitude regions.

Seed

Seed is a critical input in potato cultivation in this region as it accounts for about 40% of the total cost of production. The seed should be free from viral, fungal, bacterial diseases and nematodes. It should be procured from reliable source. Since, Ladakh has no formal systematic programme to produce quality seed in the region, procurement of the seed from the agencies situated in plains also leads to high cost of transportation to this part as roads are closed due to heavy snowfall during the planting time. Transportation of the seed by aircraft accounts heavy expenditure and is uneconomical besides the imported seed also has the problem of physiological age. To overcome this, a joint study by the scientist of DIHAR, Leh and CPRI, Shimla revealed that the region has potential to multiply breeder seed upto certified seed in the villages situated about 3600–3900 m altitude (Garg *et al.* 2003). The use of 30-45 g tuber size was the most economical for ware potato crop.

If potato requirement is calculated according to the ration scale provided to Army (140 g fresh potato/person/day or 51 kg/year) than 5100 MT potato is required for one lakh population, which can be produced from 170 ha land @ average 30 MT/ha yield. The requirement of seed potatoes will be 510 MT @ 3.0 MT/ha. For seed crop additional 18.87 ha (11-12%) land is required to produce seed of different category (Table 9.2).

Table 9.2: Seed requirement of different categories to produce ware crop for one lakh population

Grade of seed	Area requirement (ha)	Seed requirement (MT)	Production (300 MT/ha)
Breeder Seed	-	-	0.51 *
Foundation – I	0.17	0.51	5.1
Foundation – II	1.7	5.1	51
Certified Seed	17	51	510
Ware Crop	170	510	5100
Total	188.87	566.61	5666.1

* Procurement from suitable source

Because of large variation in size of the tubers, it is not possible to utilize whole produce as seed. On the other hand adjustment in time of planting, seed tuber size, spacing, manure and fertilizer doses, 75-80% produce can be obtained in term of seed, whereas < 25 g and > 60 g tubers can be utilized for table purpose. Therefore, 20-25 % area of total potato crop is needed under seed production.

In absence of infrastructure, it is difficult to produce nucleus and breeder seed of potato in this region. But the results from TPS production and utilization has shown that TPS seedling tubers used as seed gave at par yield of seed tuber crop. Therefore, use of TPS technology has a great potential in producing quality planting material in the region. It also has a added advantage of storage for long period (7-8 years), and more over the potatoes used as seed can also be utilized for table purpose. The TPS technology can be adopted using the following model:

(1) Direct TPS seeding in nursery bed

(2) Potato crop in nursery bed

(3) Harvested seedling tubers

Plate 16

True potato seed (TPS) technology model for self-sufficiency

Step I : Hybrid TPS production

1. To produce 1 kg hybrid TPS, requirement of area is as under :

   a. Requirement of parental lines

Female Lines	-	45 Kg
Male Lines	-	15 Kg
Total	-	60 Kg

   b. With 40 g of average tuber weight, area required is as under :

   i. No. of tubers present in 60 kg parental lines = 1500

   ii. Net area required to plant 1500 tubers (60 cm x 20 cm) = 180 m^2

   iii. Gross area required to produced 1.0 kg hybrid TPS = 200 m^2 (approx.)

Step II : Seedling tuber production using 10 g hybrid TPS by direct seed sowing

1. TPS (true potato seed)

10 g

2. Sowing time

Last week of May – first week of June

3. Requirement of area for sowing of 10 g TPS

80 m^2 (12.5m x 6.40m)

4. Field preparation

The nursery bed should be well prepared by 2-3 ploughings followed by planking.

5. Organic manure

400 kg well decomposed organic manure should be mixed at the time of field preparation.

6. Fertilizers

Application of 500 g Urea + 1.0 kg DAP + 250 g MOP at the time of field preparation.

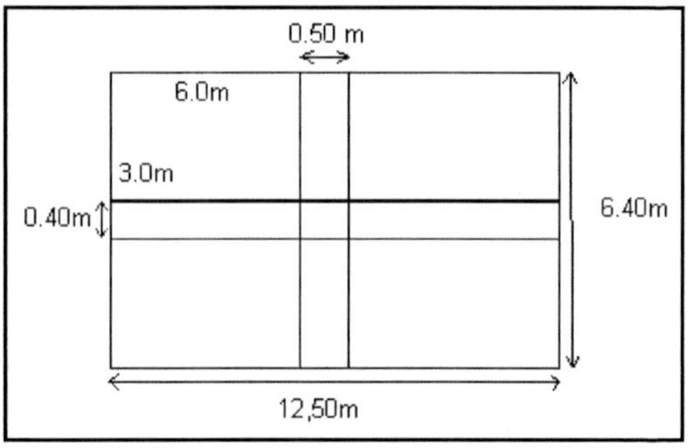

Fig. 9.1. Layout of raised beds

7. Raised bed preparations

a) 15 cm raised beds
b) 6 m x 3 m size
c) 4 numbers of beds
d) Lay out as given in Fig.9.1

8. Spacing for seed sowing

a) 2 cm deep furrow
b) 25 cm row to row distance
c) 2 cm plant to plant maintained by thinning and gap filling

9. Rows and plants/plot

24 rows/plot and 125-150 plants/row

10. Seed sowing

a) Seed should be sown in furrow such a way to put 200 seeds in a furrow. It can be achieved by mixing the TPS with sieved 500 g organic manure, before sowing.

b) After sowing of seeds, the furrow should be covered by 1:1 sieved mixture of soil and organic manure.

c) The beds should be covered by thin layer of dry grass

11. Irrigation

a) After covering by dry grass the beds should be irrigated with the help of water can with thin sprinkler.

b) Upto one week the beds should be irrigated twice/thrice in a day.

c) After 10-12 days, the dry grass should be removed from the beds.

d) After 12-15 days the beds can be irrigated by flood method as and when required.

12. Weeding, thinning and gape filling

a) The beds should be kept weed free by regular weeding.

b) At one place, one plant should be left by thinning the extra plants after 30-35 days of sowing.

c) 125-150 plants per row should be maintained by thinning/transplanting of the plants.

13. Fertilizers and earthing up

a) After establishment of the seedlings (20-25 days after sowing) 20-25g Urea/bed should be broadcasted twice at 3-4 days interval for healthy growth of the seedlings.

b) Earthing up should be done after 45 days of sowing after mixing of 25 g Urea/bed.

c) Second earthing up should be done after 60 days of sowing after mixing of 75 g Urea/bed.

14. Haulums Cutting

130 days after seed sowing

15. Harvesting

12-15 days after haulums cutting

16. Expected micro-tubers yield

a) 280 kg

b) 10,000 numbers

(1) Hybrid TPS production on commercial scale at farmer field

(2) Community water reservoir

(3) Potato field preparation before winter

Plate 17

The seedling tubers can be used as seed following standard cultural practices for ware potato crop:

Seed size (g)	Spacing Inter row (cm)	Intra row (cm)	Approx Seed rate(q)
20-40	45-50	20	30-33
10-20	45-50	15	20-22
5-10	45-50	10	15-17

Seed treatment

Boric acid is a safe and eco-friendly chemical which has replaced hazardous and highly poisonous organo-mercurials (now banned) till recently used for potato seed tuber treatment. Potato tuber borne diseases viz. black scurf (*Rizoctonia solani*), common scab (*Streptomyces scabes*), dry rots (*Fusarium* spp.), and bacterial soft rots (*Erwinia* spp.) can be controlled effectively by seed treatment with 3% boric acid. Seed potato tubers should be treated 15 days of curing in heaps of the harvest but before storage for next year planting. Following package can be followed for efficient and economical use of boric acid seed treatment for control of the tuber-borne diseases.

i. Prepare 3% boric acid by dissolving 45 g of pure boric acid (Laboratory grade 99.5% pure) in 1.5 liter of water to treat one quintal of seed potatoes by spray method. Total quantity of the chemical required for the seed to be treated should be worked out accordingly. For example to treat 100q of seed 4.5 kg of boric acid is to be dissolved in 150 liter of water.

ii. Seed potato tubers free from any soil/dirt are spreaded in a single layer over clean cemented floor or a plastic sheet. Boric acid solution should be sprayed on the seed tubers so as to wet entire surface of the tubers. Spray should be repeated till all surface of the tubers covered with the solution.

iii. Quick dip treatment (dip and remove immediately) with boric acid is equally effective but cost is almost double. In dip treatment same solution can be used up to 30 dips.

iv. The treated tubers are to be dried completely in shade before filling in gunny bags.

Cost of treatment

4.5 kg of boric acid is required to treat 100q of seed tubers. The cost of good quality (Laboratory grade 99.5% pure) boric acid comes to around Rs. 110/kg. Average cost of chemical required to treat one quintal of seed by boric acid through spray is around Rs. 5/ = as against Rs. 12 with the organo-mercurials. Spray treatment with boric acid also saves time since effective organo-mercurials treatment requires 20 minutes soaking of the tubers.

Planting time

In hills, planting commence as snow melts and soil is in workable condition. But in high altitude cold desert conditions snowfall is less in the valleys with high day temperature and low night temperature (Table 11). Therefore, the soils are in workable conditions about one month earlier than planting time. Early planting leads to damage of tender sproutings due to sudden fluctuation in temperature as heavy fluctuation in temperature is a critical feature of the regions. Although, potato is a low temperature loving crop, it requires optimum temperature (18-22° C) for veretative growth while the temperature 10-18° C is good for tuberization. In high altitude cold desert climatic conditions it grows from 5° C to 35° C with average about 55% relative humidity conditions from May-September. Planting time varies from April-June depending upon the altitude and location. The best time for planting is when stable temperature is more than 6° C during night, as 3-4° C night temperature fluctuation has been observed during April-May in the region (Table 11).

Spacing and seed rate

In remote areas of high hills, farmers plant the potato crop manually or making the furrows with bullock drawn plough and close spacing is a traditional practice. Spacing commensurate with seed size, affects the seed rate and economical return. With close spacing tillage practices can not be carried out properly. Therefore, it is desirable for planting of 30-50 g tuber size at 50-60 cm inter row and 15-20 cm intra row spacing to get maximum yield and economic return. 30-40 g TPS seedling tubers has been considered suitable for planting at 50x20 cm spacing to get good yield in the region.

Depth of planting

Planting depth depends upon type of soil, soil moisture and seed size. Deep planting in sandy or sandy loam soil assures better moisture availability during summer months. In general about 15 cm planting depth is desirable but it can differ with size of the seed. Small seed can be planted at 10 cm, medium at 10-15 cm and large at 15-20 cm depth.

Method of planting

Flat bed planting of potato with the help of Khurpi or planting in furrow is a common practice for small scale cultivation. On commercial scale, farmers make furrows (about 10 cm deep) with the help of bullock drawn plough at about 30-40 cm distance. The tubers are planted at 8-10 cm distance in the furrows followed by leveling of the field. Present study revealed that to obtain good yield, the furrows should be made at proper distance and after planting of the tuber at proper spacing in the furrows, ridges should be formed on the furrows. It helps to maintain the moisture. Planting at proper distance in rows also makes easy to carry out tillage operations. East-west row orientation and planting of TPS seedlings on the northern side of the ridge has found to be beneficial since temperatures on the northern aspect are lower than southern aspect of the ridges thus plant mortality is higher at high temperature during the period. However, north-south row orientation have a positive effect on yield in crop planted by seedling tubers, which may be attributed to better light penetration and optimum utilization of radiant energy by the crop. It also protects the soil moisture and the plants from dry and heavy wind velocity which is a common phenomena in afternoon in the region.

Manures and fertilizers

Potato crop is a most sensitive to nutrient stress as it has shallow and sparse root system. The organic manures not only supply the nutrients but also improve the physico-chemical properties of the soil for better plant and tuber growth. Fertilizer management based on soil and plant test not only maintain soil-ecosystem, but also raises potato productivity.

For optimum yield, the potato plant requires a sufficient amount of nitrogen along with adequate supplies of phosphorus, potassium and the other nutrients i.e. calcium, magnesium, sulphur and micronutrients i.e. iron, zinc, manganese, copper, boron and molybdenum. Many of these micro-nutrients play a vital role in metabolic enzymes functioning. Even though these elements are needed only in traces, many of cold desert soils may not supply them in sufficient quantity for proper growth and optimum yield of potato. Hence, the deficiency symptoms in potato crop have been observed in many parts in the region.

Nitrogen is the first limiting nutrient for potato production whose availablity is low in sandy alkaline soils of the high altitude regions. Its deficiency shows stunted growth and the number and size of the tubers as well as bulking period is reduced. Phosphorus enters into the structural composition and increases number of medium size tubers. Potassium on the other hand increases the leaf area and its duration and helps in utilizing nitrogen more efficiently and increases tuber size and hence yield. Besides imparting resistance against diseases and water stress, it also provides frost resistance to the plant (Saini, 1978; Grewal and Singh, 1980).

The yield obtained with combined use of organic manure and fertilizers was higher than fertilizers alone. To harvest optimum yield, depending upon moisture content, 20-30 tones per hectare of well decomposed organic manure should be mixed well in the field. Decomposition of the FYM is a major problem in the region because of low temperature. Therefore, at least one year old well decomposed FYM should be used for better release of the nutrients to the soil. The more yield with combined use of organic manure and fertilizers has also been observed in other crops (Singh and Singh, 2003). In addition to the organic manure, application of N 100 kg/ha, P_2O_5 60 kg/ha and K_2O 60 kg/ha is desirable to obtained good yield of potato crop in the region. Availability of the fertilizers is also difficult but some sources for N, P_2O_5 and K_2O nutrients are commonly available in the region (Table 9.3).

Table 9.3. Commonly available fertilizers in the region

Nutrients	Source	Composition (%)		
		N	P_2O_5	K_2O
Nitrogen	Urea	46	-	-
Phosphorus	Diammonium phosphate (DAP)	18	46	-
Potassium	Muriate of potash (MOP)	-	-	60

The above mentioned sources are normally available in 50 kg packing which contains urea (23 kg N), DAP (9 kg N and 23 kg P_2O_5) and MOP (30 kg K_2O). Some fields are deficient in zinc, iron, boron, manganese, copper, molybdenum which require additional application of the micro-nutrients. These nutrients are required in low quantity and can be applied as per the detail given in Table 9.4.

Table 9.4: Doses of micro nutrients application in potato crop

Nutrients	Source	Soil application (kg/ha)	Spray application (g/100 litter water)
Zinc	Zinc sulphate	25	200
Iron	Ferous sulphate	25	200
Manganese	Manganese sulphate	25	200
Copper	Copper sulphate	10	100
Molybdenum	Ammonium molybdate	1	50
Boron	Sodium borate	1	50

Depending upon the altitude, the organic manure should spread before ploughing in the field from April-May. It should be mixed well in the field at the time of ploughing. After that a pre–irrigation should be given to the field. At proper moisture the field should be ploughed after broadcasting of the half dose of N and full dose of P and K followed by planking. Just after the preparation of the field, planting should be done with in two to three days 50% to avoid moisture loss from the field. The remaining half dose of nitrogen can be applied twice, in 40 and 60 days of planting at proper moisture level.

Irrigation

Management of soil moisture is essential for optimum yield under assured irrigation. Water from snow melt, rivers and natural springs are used for irrigation by gravitational system. This water either harvested in community reservoirs first (Fig. 9.2.2) or used directly to the fields. High wind velocity and low atmospheric humidity lead to more evapo-transpiration which increases the water requirement of potato crop in the region. Mulching with 200 micron thick black polythene sheet saved about 30% water to the crop. Sandy soils required comparatively 20-30% higher number of irrigations. Pre-planting irrigation is beneficial for uniform emergence and also increase the duration of first irrigation after planting. Normally, first light irrigation is given 20-25 days after planting and subsequent irrigations at 12-20 days interval depending upon the type of soil and weather conditions prevailing in the area. During the irrigation, water should not be reach more than 2/3 height of the ridges.

Cultural practices

Timely control of the weeds in potato crop is essential as they compete for nutrients, moisture, space and light and consequently reduce the crop yield. Well preparation of land is beneficial as it delays the germination and growth of the weeds. Planting of potato crop at proper spacing in rows makes it easy to remove the weeds and earthing up. One fine tillage operation should be carried out after first irrigation about 25 days of planting and second tillage may be done at 20-30 days of first tillage operation depending upon weeds population. At the time of second cultural operations weeding, hoeing, fertilizer application and earthing up are carried out at a time. The earthing up is also helpful to avoid greening of the tubers. Weed population is low when the crop covers the soil by spreading of the canopy of the plants and these thinly populated weeds can be uprooted by hand.

Soil solarization

Soil solarization is the technique to treat the field soil by trapping solar heat during hot summer months, by using transparent polythene sheet. It is effective, simple, physical and non hazardous eco-frendly method to control soil born diseases, pests and weeds. Heated water vapors trapped below polythene sheet, increase thermal conductivity and the heat is retained in soil for much longer period causing an effective control of the soil-borne pathogens. In a trial conducted on solarization for sixty days (15 April to 15 June in 2001) at Defence Institute of High Altitude Research, Station Ranbirpur at 3600 m altitude resulted 60% reduction in weed population in comparison to control. It reduced the efforts during tillage operation and save on the nutrients which would have absorbed by the weeds. There was 90% reduction in the injured potato tubers by the cutworm which is a common problem in the region.

Harvesting

Keeping in view the demand and price in the market, potato crop can be harvested early in the month of August at slightly immature stage specialy below 3000 m altitude. But these potatoes are not suitable for long duration storage. To maintain the supply of locally produced potato for longer duration, more than 80% produce must be stored during long freezing winter season. For storage, potato crop should be harvested at proper maturity for which irrigation should be stopped well before. Haulms must be removed close to the ground level and regrowth of stumps be checked. The crop is harvested when the soil is in workable condition and the tuber skin is hardened. It can be achieved by harvesting the tubers 12-15 days after haulm cutting. The produce should be kept in shady place for further curing.

Crop protection

Major diseases of potato i.e. late blight, early blight, black scurf, wart are not a serious problem in high altitude cold desert region. This may be due to long prevailing severe winter which is not conducive to survive/spread of many pathogens and insect-pests. However, potato crop in the cold desert is affected to some extent by insect-pests and disease. Therefore, health management practices are required for conrol of common scab, cutworm and aphids which affect the health of potato crop in the region. These disease and insects-pests were observed to prevail up to 3600 m AMSL. The intensity of the diseases and insects decreases with the increasing of altitude (Garg *et al.* 2003). The details of the common disease and insect-pests and their control measures are given below:

Common Scab (*Streptomyces* Spp.)

Common scab is a wide spread disease in almost all potato growing regions specially in lower areas of high altitude (upto 3600 m). The pathogen mainly affects potato tubers. Small redish or brownish spots on tuber surface which enlarge into circular or irregular lesions with sunken corky spot or pitted lesions having deep star shaped cracks or russetting of tuber surface. The disease normally does not cause yield losses but disfigures the tubers (Plate 18.1), thereby reducing the market value and increasing peeling losses. The infected tubers are unfit to use of seed. Severely infected tubers are also not suitable for storage during winter season because of heavy weight loss and spoilage during storage. Fungal infection was also observed in presence of high humidity during storage on tubers affected by common scab.

Control

1. Use only disease free seed tubers.
2. The pathogens of the disease survive for long duration in the soil, therefore, 3-4 years crop rotation with wheat, barley, pea should be followed.
3. Give tuber treatment with 3% boric acid before storage and dry in shade before storage or planting.
4. Maintain high moisture in the field at least for a few weeks during the initial tuberization phase (July).
5. Plough the field during October/November and leave the soil exposed to low temperature during winter months.

Cutworms (*Agrotis* Spp.)

The caterpillars (Plate 18.2) cause crop damage. They feed at night on young shoots or under ground tubers. In early stage of the crop, the caterpillars cut the stems of the young plants

(1) Tubers of Sindhu Red affected by common scab

(2) Tubers of Kufri Chandramukhi affected by common scab

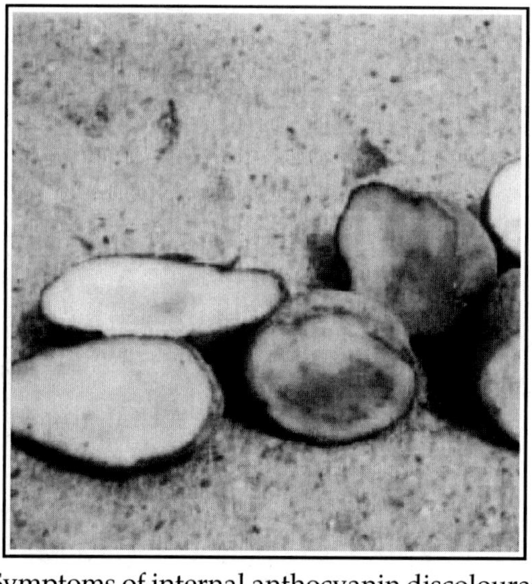

(3) Symptoms of internal anthocyanin discolouration

Plate 18

(1) Potato tubers damage by cutworm

(2) Caterpillar of potato cutworm

(3) Potato aphids

Plate 19

just above the ground and feed on shoots and leaves. During later stage, they feed on tubers by boring and nibbling (Plate 19.1) into them which ultimately affects the yield and market value of the crop. In highly infested fields in lower altitude regions, as high as 30% tubers are observed to be damaged by cutworms. The drought condition favours high incidence of cut warms. The caterpillar feeds on several host crops i.e. carrot, radish, cabbage, onion.

Control

The insect can be effectively controlled by using integrated control measures as:

1. Drench the ridges with chlorpyriphos 20 EC @ 2.5 l/ha.
2. Apply thimet @ 10 kg/ha at planting and 10 kg/ha at earthing up.
3. Adopt 2-3 years crop rotation.
4. Ploughing of the field during winter months.

Aphid (*Myzus persicae & Aphis gossypii*)

The aphids although directly do not affect the crop but pose a serious problem by acting as vector for viruses in the crop. The aphid population has a major role to produce healthy seed potatoes. In a study, 40 aphids per 100 compound leaves have been counted at 3300 m altitude, 21 at 3600 m and the aphid population at 4000 m altitude was recorded nil in potato growing villages in Ladakh (Garg *et al.,* 2003). It is a greenish yellow, green or sometimes pinkish with a pair of well developed head frontal tubercles. Winged aphid has blackish head and greenish abdomen with dark patch on dorsal surface. Adult and nymphs suck the sap from foliage and leaves curl downwards (Plate 19.3). If the population is very high, infected leaves turn yellow and die. However, the main damage caused by these insects to the potato production in lower altitude (below 3300 m) is by the dissemination of viral diseases.

Control

1. Application of thimet or phorate @ 10 kg/ha at planting.
2. Spray the crop in July/August (at the appearance of the aphid) with thiomethoxam WG @ 100 g/ha.

Internal anthocyanin discolouration

Internal anthocyanin discolouration is an internally visible disorder (Fig. 9.4.3). When a tuber affected by internal anthocyanin discolouration is cut, it show, purplish red stripes and flame shaped discolouration due to the formation of anthocyanin in the cell sap. Light appears to play a role in the formation of this discolouration, which is mainly observed in green tubers, especially in the green part of such tubers. The discolouration becomes a little less apparent when the tubers are cooked, but it nevertheless affect the quality of ware potatoes. To control this disorder, the tubers should not exposed to sunlight during the cropping period.

To produce healthy potatoes, the various integrated approaches should be followed to control the insects-pests and diseases, as under:

1. Use of clean, healthy and disease free seed.
2. Seed should be planted at proper time on the basis of altitude.
3. The soil should be well prepared which bring the soil into better physical condition for better emergence and proper plant growth.
4. Plough the soil during October/November just before onset of severe winter. The soil should remain open for a long winter season which will be helpful in destroying

the photogene and protective stage of the insect pests (pupa) due to sever cold.

5. The seed should be used of right physiological condition.
6. Maintain proper moisture in the field specially at early tuberization stage in the month of July.
7. Regular watching and rouging of infected plants from the field.
8. Timely use of fungicides or insecticides just after appearance of the initial symptoms.
9. Adopt 2-3 years crop rotation to avoid build up of pathogens and parasites.
10. Crop like wheat, barley, peas and oats should be included in the rotation.
11. Isolation should be maintained at least of 25 m between seed and ware crop.
12. The crop should be harvested at proper time to avoid frost injury in the tubers.

Post–harvest handling and storage

Harvesting time in the region depends upon altitude and planting time. In most of the areas potato crop is harvested in September when weather is dry and cool. At harvest, tubers are liable to considerable skinning, bruising and other damages, which subsequently lead to rottage if not handled properly. In the extremely harsh climatic conditions traditionally farmers store potatoes in 1.2–2.1 meter deep underground pits. The diameter or length–width of the pit depends upon the quantity to be stored. After harvesting in October the potatoes are put into this pit and covered with 45-60 cm dry soil. Following disadvantages have been observed in this method of potato storage known in the local art:

a. The required quantity at the desired time cannot be taken out from this type of pit storage. After opening the pit we have to take out all the potatoes from the pit to avoid the spoilage especially during severe winter months.

b. Heavy spoilage in the tubers is found due to inadequate temperature and humidity in the pit storage.

c. Potato tubers can be stored only for limited period (3-4 months).

d. The tubers are directly in contact with soil therefore, the peel and colour are affected badly.

e. Due to rough appearance farmers get lower price of their produce in the market.

f. Microbial infection to the tubers result decomposition of the tubers.

g. In this method sprouting starts in the tubers in the month of March and it become difficult to further store the tuber.

h. Cent percent tubers have shriveling effect after the month of April.

i. Heavy weight loss, sprouting and shriveling affected tubers are not suitable for consumption.

Handling at harvest

For lone-term storage, potato crop should be harvested at proper maturity stage. At reaching the maturity stage, irrigation should be stopped well before. Haulms should be removed close to the ground level and regrowth of stumps be checked. The crop should dugged out when the soil is in workable condition and the tuber skin is hardened. This could be achieved when tubers are harvested 12-15 days after haulms cutting.

Sorting

In general, potato crop is harvested in September- October depending upon altitude in this region. At that time, weather is dry and cool, and tubers are liable to considerable skinning, bruising and other damages, which further lead to the rotting. The immature tubers are also not suitable to store for longer period. Therefore, all type of damaged and less than 25 g tubers should be sorted out after harvesting the crop.

Curing

The tubers harvested from fully matured crop should be kept in heaps at any ventilated place or the heaps can be made on the dry soil in the field. The heaps in the open field should be covered with a thick layer of straw or gunny bags to protect the tubers from severe cold and direct sunlight and to maintain an equitable temperature with in the heap. In the heaps, potato tubers are cured for 8-12 days under the dry weather conditions of the region. This procedure heals the shallow wounds, hardened the skin and makes the tubers free from soil.

Packing

After curing sorting is necessary to remove diseased, damaged, under size and other undesirable tubers. After curing and sorting out, the potatoes should be filled up in nylon netted bag (leno bags). Packing of 50 kg potatoes in a 105 cm length and 55 cm width size bag is suitable to stacked on the cemented strips in the store and easy to handle. 0.25 mm thick bags with pore size about 5 x 5 mm are found suitable to allow sufficient quantity of CIPC to enter inside the bags for deposition on surface of the tubers.

Layout of the bags

The bags should systematically be stacked on the cemented strip in layers inside the underground store and arranged with sufficient space between the lines of bags to allow circulation of CIPC fog, respiratory air and heat of radiation omitted by earth, so as to maintained moderated temperature within in structure. About 0.30 m empty space is kept as a clearance between roof and the top of the stacked bags.

Treatment of CIPC

To check the sprouting, tubers should be treated with CIPC chemical. Common name of CIPC is chlorpropham and chemical name Isopropyl 3 – chlorocarbanilate. The fog of the chemical is applied with a fogging machine. LPG based handy fogging machine found suitable to use in cold desert high altitude conditions of Ladakh. The CIPC chemical is injected as fog in the store through fogging inlet pipe.

i) Time of treatment

Fogging of CIPC chemical at above rate should be done twice, the first in first fortnight of October and second in first fortnight of March. The $1^{\circ} \pm 2^{\circ}$ C temperature of store is optimum at time of fogging.

ii) Post-treatment activities

a. During and after fogging, the store should remain airtight for 48-72 hours for fog vapors to settle down on the tubers.

b. After 48-72 hours of fogging, the store should be open at mid daytime to remove the excess carbondioxide from the store.

c. Nobody should enter immediately after opening the store, as the person can go under unconsciousness stage due to high level of corbondioxide inside the store.

Technology dissemination

Four trials have been conducted on different aspects of long –term potato storage at Defence Institute of High Altitude Research. This technology has been demonstrated to the farmers, officials of State Agriculture Department, Farmer's Co-operative Marketing Society and NGOs and these organizations can play an important role in further dissemination of this technology in far-flung villages of Ladakh. It will be helpful to make the region self sufficient in potato production and round the year availability to the army deployed in the region and civil population. The farmers of the region will be able to use their own crop produce throughout the year and it will give then remuneration.

Socio-economic consideration

More than 90 percent population is associated or depended upon agriculture for their livelihood in cold desert regions. The short duration crop of potato can play an important role for food and nutrition security and can increase their farm income by producing large quantity of potato to fulfill round the year requirement of civil market and army deployed in the region. At present farmers are supplying about 1500 MT potatoes per annum to the army and getting more than Rs 150.0 Lakh per annum. 100% requirement of armed forces can be meet out with the adaptation of technology for long-term potato storage with CIPC treatment and the farmers can earn more income.

Future strategies

The Defence Institute of High Altitude Research has developed the process for long-term potato storage on commercial scale. With adaptation of this technology self-sustainability can be achieved for potato production and round the year availability of local produce. This process has three main components i.e. specially designed underground store, specific post-harvest activities up to the loading of store and fogging treatment of CIPC. There is need of petrol/diesel/LPG based fogging machine. For fogging treatment petrol/diesel based fogging machine do not work properly may be due to the availability of less oxygen. But LPG based fogging machine was found suitable. Farmers who want to store potato with this technology require to arrange the costly fogging machine. Therefore, simultaneously DIHAR is also working to standardize the technique for storage the potatoes with ordinary spray treatment of CIPC and it will be a house holds technique of potato storage for any quantity.

□□□

CHAPTER **10**

Potato Sustainability

To make the region self sufficient, efforts were made to gather scientific information on important aspects of potato production in high altitude cold desert climatic condition. This information is helpful in boosting the production and increase the availability period of locally produced potato. But there is further need to evolve the techniques keeping in view the physiographic conditions of the region. Varietal trials conducted to find out suitable variety and identified Kufri Chandramukhi as a suitable cultivar for the region. Production can be increased with the adoption of cultural practices developed on scientific basis for the region and described in this manuscript. Availability of water for irrigation is a major constraint, beside in early stage of potato planting, low soil temperature has an adverse affect on emergence, proper growth and development of plant. Mulching of black polythene (200 micron thick) conserve not only the soil moisture but reduces about 30% water requirement of potato crop. In addition, in early stage of crop growth when temperature is low it helps in increasing the soil temperature by 2-6° C. Mulching suppresses the growth of weeds and ultimately increases the tuber yield.

Use of healthy seed is important because of the prevalence of a number of viral, fungal and bacterial diseases, which reduce the yield considerably. The bulky nature of potato seed results in higher seed rate and transportation cost, which accounts for high cost of cultivation. Insufficient availability of quality seed is a major problem in potato cultivation in cold desert region. With out proper infrastructure it is difficult to produce quality seed. But as described earlier, breeders' seed can be multiplied for Foundation I, Foundation II and Certified seed under proper management practices specially in the identified locations which will protect the crop from degenerative diseases during each multiplication.

On the basis of research findings, TPS technology is expected to overcome the problems of seed tubers and can supplement the availability of healthy propagules under photoperiods of 14–15 hour optimal for maximizing flowering and berry set. Bright sunshine, low rainfall during hybridization and low incidence of diseases specially late blight are favorable conditions for superior quality TPS production in high altitude cold desert regions. The crop can be successfully taken from seedling transplantation and the higher production can be achieved by producing seedling tubers in nursery bed and storing during freezing winter months with the underground storage technique which can be later on used as seed tubers in next year for potato production. For potato sustainability in the region, round the year availability of locally produce can be maintained by (i) early crop production in greenhouse (trench) and (ii) long term potato storage with CIPC treatment.

(i) Early crop production in greenhouse

During September-October fresh potato is available whereas during November-May it is available from storage of the crop in an underground store. To fulfill the requirement from June-August an early off season crop can be grown successfully on commercial level in semi-underground greenhouse. 40-60 kg potatoes can be harvested from 10x3x1 m size greenhouse depending upon harvesting time. About 1258 MT potatoes are required from June to August for one lakh population, which can be easily produced from 25150 number of greenhouses of above mentioned size @ average 50 kg/greenhouse yield. Production of early potato in the greenhouses is also beneficial in providing employment by way of construction of the greenhouses and potato cultivation during off season which is remunerative to the farmers considerably. The Fig. 10.1 reveraled that out of 100% (round the year) requirement, 16.6% can be provided in September-October from the ordinary storage of freshly harvested crop. After that 58.4% demand from November-May can be met out from specially designed underground store and remaining 25% requirement from June-August can fulfil by cultivating early (off season) potato crop in semi underground greenhoues.

(ii) Storage with CIPC treatment

As mentioned above to fulfil about 16.6% demand, potato is available from September-October from freshly harvested crop and to maintain year the round supply, it has to be stored from November-August. Therefore, 83.4% produce needs to be stored in order to maintain its availability for next 10 months (November-August). Out of 83.4% requirement 41.8% (from November-March) can be supply from the underground store and remaining 41.6% demand can be met out by the storing of potatoes with CIPC treatment (Fig. 10.2).

YEAR THE ROUND AVAILABILITY OF LOCALLY PRODUCED POTATO

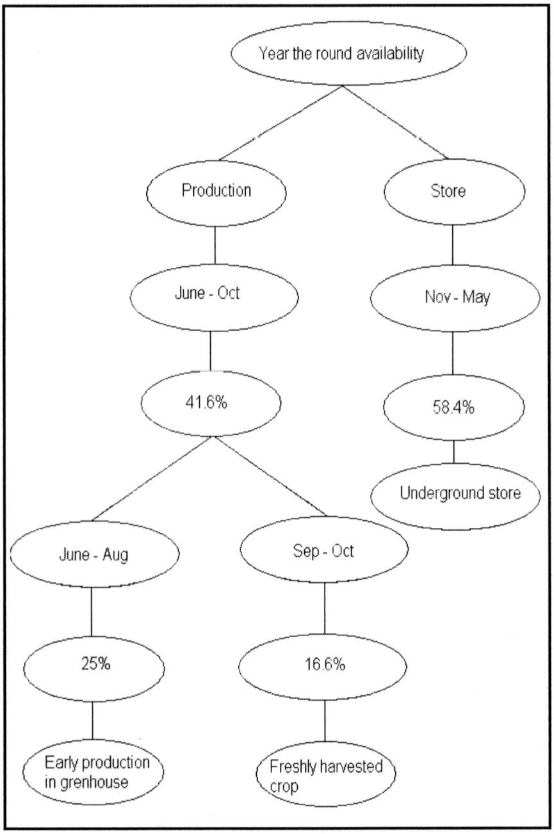

Fig 10.1. Early crop production

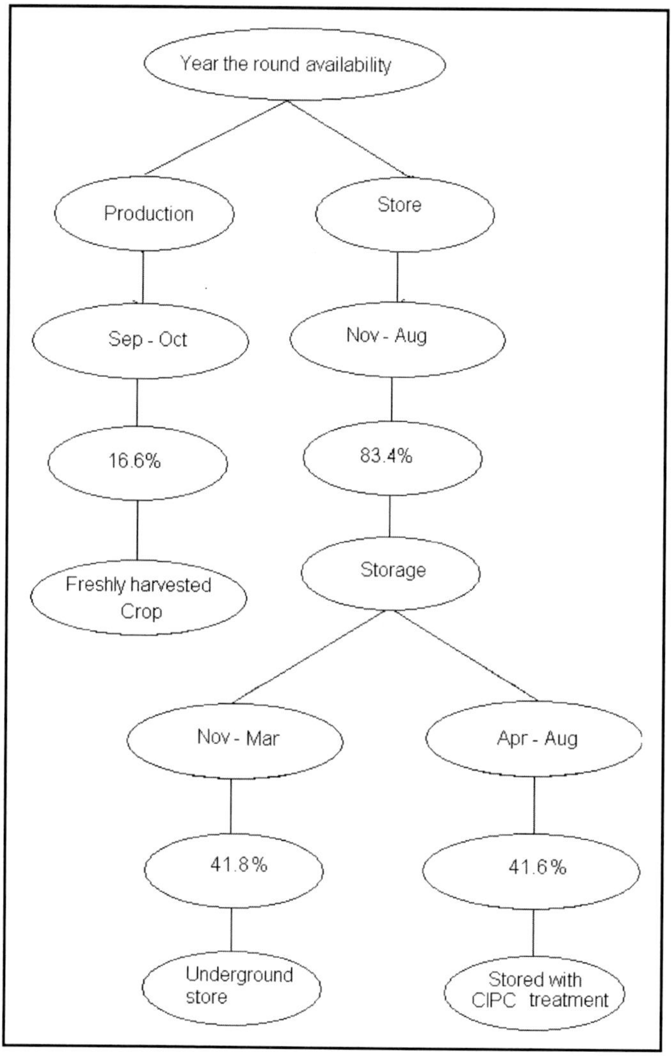

Fig 10.2. Storage with CIPC treatment

❑❑❑

CHAPTER **11**

Weather Data of Cold Desert

Table 11 : Temperature and humidity data of Leh (3450 m altitude)

Table 11.1 (January)

	Minimum (°C)			Maximum (°C)			Humidity (%)		
Date	2005	2006	Average	2005	2006	Average	2005	2006	Average
1	-6.9	-6.6	-6.75	-1.0	0.5	-0.25	85	52	68.5
2	-4.6	-5.5	-5.05	2.6	-2.4	0.1	68	58	63
3	-5.9	-6.8	-6.35	1.7	-0.5	0.6	46	81	63.5
4	-12.5	-14.4	-13.45	1.7	-4.9	-1.6	72	80	76
5	-11.0	-15.9	-13.45	-2.3	-5.9	-4.1	77	82	79.5
6	-10.9	-16.4	-13.65	-1.8	-0.6	-1.2	78	76	77
7	-14.5	-17.8	-16.15	-0.5	-1.5	-1.0	75	79	77
8	-13.7	-17.3	-15.5	0.2	-1.3	-0.55	66	69	67.5
9	-13.7	-15.0	-14.35	0.3	-1.5	-0.6	64	75	69.5
10	-12.8	-14.1	-13.45	0.0	-3.4	-1.7	69	68	68.5
11	-14.9	-11.5	-13.2	-1.9	-0.2	-1.05	71	77	74
12	-14.8	-11.7	-13.25	-1.8	1.7	-0.05	57	72	64.5
13	-11.2	-12.6	-11.9	-0.8	0.1	-0.35	60	70	65
14	-11.3	-10.2	-10.75	1.6	0.6	1.1	65	72	68.5
15	-13.9	-9.2	-11.55	4.3	0.1	2.2	59	74	66.5
16	-15.4	-5.8	-10.6	0.9	-2.1	-0.6	51	85	68
17	-14.7	-7.0	-10.85	1.5	-1.5	0.0	69	87	78
18	-14.9	-7.3	-11.1	-1.3	-4.0	-2.65	70	83	76.5
19	-14.1	-14.2	-14.15	-4.0	-1.0	-2.5	70	82	76
20	-16.0	-16.1	-16.05	-3.4	-2.5	-2.95	67	80	73.5
21	-10.4	-15.8	-13.1	-3.0	-6.8	-4.9	69	74	71.5
22	-15.3	-18.9	-15.1	-1.9	-2.9	-2.4	54	68	61
23	-10.0	-17.4	-13.7	-5.0	-8.4	-6.7	65	82	73.5
24	-9.4	-18.9	-14.15	-3.0	-7.0	-5	59	84	82.5
25	-11.4	-22.0	-16.7	-0.4	-5.5	-2.95	57	79	79.5
26	-15.3	-22.5	-18.9	-3.5	-7.7	-5.6	58	78	60
27	-14.7	-20.9	-17.8	-4.2	-7.7	-5.95	69	70	59
28	-13.0	-20.9	-16.95	-3.3	-6.7	-5	66	74	60.5
29	-16.9	-20.6	-18.75	-4.0	-4.4	-4.2	64	80	68
30	-12.0	-11.2	-11.6	-5.7	-1.9	-3.8	73	72	65.5
31	-15.1	-10.5	-12.8	-2.0	-2.0	-2	62	80	68
Average	**-10.29**	**-14.03**	**-13.26**	**-1.29**	**-2.94**	**-2.12**	**65.64**	**75.58**	**69.98**

Table 11.2: (February)

	Minimum (°C)			Maximum (°C)			Humidity (%)		
Date	**2005**	**2006**	**Average**	**2005**	**2006**	**Average**	**2005**	**2006**	**Average**
1	-15.4	-7.5	-11.45	-2.9	-0.1	-1.5	74	76	75.0
2	-15.5	-12.4	-13.95	-1.3	1.9	0.3	67	82	74.5
3	-13.6	-12.2	-12.9	-1.2	2.2	0.5	58	70	64.0
4	-10.7	-14.3	-12.5	0.3	0.8	0.55	69	84	76.5
5	-9.4	-13.2	-11.3	0.1	2.0	1.05	66	84	75.0
6	-12.7	-14.4	-13.55	-1.3	2.2	0.45	65	78	71.5
7	-9.6	-10.1	-9.85	1.9	1.6	1.75	78	76	77.0
8	-6.8	-8.4	-7.6	-1.5	2.2	0.35	83	76	79.5
9	-10.5	-7.4	-8.95	-1.7	3.2	0.75	80	73	76.5
10	-7.9	-11.0	-9.45	-1.9	2.1	0.1	81	76	78.5
11	-7.0	-9.2	-8.1	-1.5	3.2	0.85	82	80	81.0
12	-5.7	-7.2	-6.45	-0.7	3.3	1.3	81	81	81.0
13	-5.7	-2.6	-4.15	-0.5	3.8	1.65	73	72	72.5
14	-8.3	-2.7	-5.5	0.0	5.1	2.55	65	72	68.5
15	-7.7	-2.1	-4.9	-0.8	4.8	2.0	84	81	82.5
16	-9.8	-5.9	-7.85	-2.9	5.0	1.05	76	72	74.0
17	-15.5	-3.7	-9.6	0.1	3.9	2.0	68	81	74.5
18	-11.1	-3.7	-7.4	-0.4	4.9	2.25	58	74	66.0
19	-5.0	-2.7	-3.85	1.2	8.3	4.75	46	74	60.0
20	-5.3	-0.4	-2.85	1.7	7.6	4.65	44	83	63.5
21	-12.7	-4.7	-8.7	5.0	7.6	6.3	64	53	58.5
22	-10.5	-4.7	-7.6	2.8	7.6	5.2	69	66	67.5
23	-11.9	-4.7	-8.3	0.3	7.6	3.95	66	63	64.5
24	-7.8	-4.5	-6.15	0.8	7.8	4.3	79	73	76.0
25	-6.8	-1.7	-4.25	0.4	8.4	4.4	75	55	65.0
26	-7.8	-1.4	-4.6	1.9	9.2	5.55	60	84	72.0
27	-3.3	-1.7	-2.5	2.7	5.6	4.15	60	74	67.0
28	-3.0	-7.2	-5.1	-2.8	4.7	0.95	80	62	71.0
Average	**-9.17**	**-6.48**	**-7.83**	**-0.07**	**4.51**	**2.22**	**69.67**	**73.46**	**71.89**

Table 11.3: (March)

	Minimum (°C)			Maximum (°C)			Humidity (%)		
Date	2005	2006	Average	2005	2006	Average	2005	2006	Average
1	-3.5	-8.0	-5.75	3.9	3.9	3.9	68	55	61.5
2	-2.9	-8.0	-5.45	4.7	5.6	5.15	63	56	59.5
3	-11.4	-5.3	-8.35	5.3	5.8	5.55	52	54	53.0
4	-2.7	-6.1	-4.4	7.0	6.4	6.7	51	56	53.5
5	-3.4	-5.2	-4.3	6.9	5.8	6.35	48	55	51.5
6	-3.5	-7.7	-5.6	5.8	5.8	5.8	50	54	52.0
7	5.1	-3.1	1.0	5.5	4.8	5.15	31	61	46.0
8	-4.7	-7.7	-6.2	6.0	6.4	6.2	30	54	42.0
9	-5.9	-4.9	-5.4	7.4	6.4	6.9	25	53	39.0
10	-1.5	-3.3	-2.4	8.1	6.0	7.05	40	68	54.0
11	-4.1	-3.1	-3.6	8.4	2.7	5.55	50	64	57.0
12	-4.9	-6.9	-5.9	8.2	6.4	7.3	46	52	49.0
13	-3.9	-6.7	-5.3	9.0	8.2	8.6	42	55	48.5
14	-3.7	-1.8	-2.75	11.9	8.0	9.95	41	55	48.0
15	2.2	-1.7	0.25	12.9	7.8	10.35	39	74	56.5
16	2.2	-4.1	-0.95	13.7	5.9	9.8	39	45	42.0
17	4.5	-9.1	-2.3	13.9	2.5	8.2	32	53	42.5
18	1.1	-8.5	-3.7	11.7	4.7	8.2	44	50	47.0
19	0.4	-6.9	-3.25	11.5	6.4	8.95	64	49	56.5
20	-1.7	-7.3	-4.5	9.6	7.9	8.75	70	59	64.5
21	-0.8	-3.2	-2.0	6.9	7.1	7.0	47	49	48.0
22	-1.3	-7.2	-4.25	5.0	8.2	6.6	71	53	62.0
23	-1.4	-5.3	-3.35	5.6	8.5	7.05	64	51	57.5
24	-4.3	-5.0	-4.65	4.4	7.9	6.15	55	56	55.5
25	-3.9	-4.1	-4.0	6.9	5.8	6.35	46	76	61.0
26	-1.5	-0.5	-1.0	6.8	8.9	7.85	56	59	57.5
27	-1.7	-3.9	-2.8	10.0	9.0	9.5	37	57	47.0
28	-4.9	-1.6	-3.25	7.7	9.3	8.5	54	53	53.5
29	-4.9	-3.6	-4.25	7.6	8.5	8.05	52	51	51.5
30	-1.4	-3.1	-2.25	8.2	9.4	8.8	34	51	42.5
31	-6.9	-0.4	-3.65	3.6	10.7	7.15	40	50	45.0
Average	**-2.42**	**-4.72**	**-3.36**	**7.87**	**5.87**	**6.56**	**47.77**	**55.74**	**47.27**

Table 11.4: (April)

	Minimum (°C)			Maximum (°C)			Humidity (%)		
Date	2005	2006	Average	2005	2006	Average	2005	2006	Average
1	-5.2	-1.0	-3.1	7.4	11.6	9.5	49	61	55.0
2	-3.7	3.0	-0.35	8.4	12.9	10.65	46	52	49.0
3	-4.1	0.5	-1.8	10.3	10.2	10.25	45	56	50.5
4	-3.1	1.5	-0.8	10.4	9.7	10.05	44	59	51.5
5	2.5	-2.4	0.05	13.2	10.8	12.0	51	50	50.5
6	5.5	-1.6	1.95	14.7	11.9	13.3	45	49	47.0
7	-0.6	-1.0	-0.8	10.6	12.1	11.35	87	50	68.5
8	-2.5	-0.5	-1.5	10.4	12.8	11.6	78	46	62.0
9	-2.0	4.6	1.3	4.1	13.6	8.85	57	58	57.5
10	-5.7	0.2	-2.75	1.8	12.6	7.2	64	80	72.0
11	-5.2	-1.3	-3.25	1.6	7.4	4.5	58	50	54.0
12	-5.1	-0.2	-2.65	2.9	7.3	5.1	52	38	45.0
13	-1.9	-3.6	-2.75	6.8	6.9	6.85	57	53	55.0
14	1.5	-5.0	-1.75	9.6	7.8	8.7	56	55	55.5
15	1.0	2.3	1.65	11.6	9.0	10.3	53	49	51.0
16	-1.4	0.3	-0.55	9.6	10.9	10.25	58	62	60.0
17	-1.6	-0.2	-0.9	9.8	10.2	10.0	56	65	60.5
18	-1.9	1.2	-0.35	10.8	8.5	9.65	59	68	63.5
19	-1.4	1.5	0.05	11.4	8.1	9.75	41	56	48.5
20	-0.1	0.6	0.25	12.9	10.3	11.6	45	52	48.5
21	1.8	-0.3	0.75	14.5	12.4	13.45	53	50	51.5
22	2.2	4.3	3.25	14.2	12.4	13.3	48	52	50.0
23	4.8	1.5	3.15	14.3	14.1	14.2	55	46	50.5
24	4.5	1.0	2.75	16.3	13.9	15.1	56	52	54.0
25	3.9	3.1	3.5	12.9	15.8	14.35	68	47	57.5
26	4.2	3.0	3.6	15.1	15.7	15.4	43	45	44.0
27	2.5	3.7	3.1	16.5	17.2	16.85	54	44	49.0
28	3.9	3.7	3.8	14.4	17.5	15.95	56	40	48.0
29	1.7	3.6	2.65	9.5	17.9	13.7	66	45	55.5
30	3.5	4.1	3.8	13.9	17.6	15.75	62	47	54.5
Average	**-0.07**	**0.89**	**0.41**	**10.66**	**11.97**	**11.31**	**55.4**	**52.56**	**53.98**

Table 11.5: (May)

	Minimum (°C)			Maximum (°C)			Humidity (%)		
Date	2005	2006	Average	2005	2006	Average	2005	2006	Average
1	3.6	4.0	3.8	14.4	17.2	15.8	56	69	62.5
2	4.4	5.9	5.15	16.5	13.1	14.8	64	60	62.0
3	4.5	7.2	5.85	15.1	16.5	15.8	56	54	55.0
4	5.6	1.4	3.5	16.8	13.1	14.95	66	58	62.0
5	4.0	4.9	4.45	12.2	18.4	15.3	64	44	54.0
6	3.8	8.0	5.9	15.6	19.9	17.75	61	48	54.5
7	3.6	6.4	5.0	13.9	17.4	15.65	56	46	51.0
8	3.7	7.6	5.65	16.9	18.3	17.6	44	52	48.0
9	4.2	5.2	4.7	15.4	17.8	16.6	51	69	60.0
10	3.5	5.3	4.4	14.6	18.3	16.45	62	70	66.0
11	4.5	5.5	5.0	11.4	18.3	14.85	52	49	50.5
12	3.5	7.1	5.3	15.1	18.3	16.7	50	46	48.0
13	2.3	3.8	3.05	13.6	18.9	16.25	47	35	41.0
14	1.9	7.2	4.55	12.0	20.7	16.35	62	47	54.5
15	2.3	7.1	4.7	12.9	21.6	17.25	57	44	50.5
16	5.7	7.2	6.45	14.6	22.3	18.45	50	43	46.5
17	2.6	9.8	6.2	16.3	23.4	19.85	65	45	55.0
18	4.2	8.9	6.55	14.8	23.5	19.15	60	54	57.0
19	4.5	7.1	5.8	14.9	20.0	17.45	57	56	56.5
20	5.8	8.2	7.0	14.3	21.6	17.95	48	55	51.5
21	3.9	8.5	6.2	12.8	19.5	16.15	61	50	55.5
22	1.8	11.8	6.8	8.5	21.3	14.9	60	52	56.0
23	5.5	9.1	7.3	13.5	22.0	17.75	49	47	48.0
24	3.6	8.0	5.8	16.2	22.4	19.3	58	48	53.0
25	3.4	7.8	5.6	15.1	22.8	18.95	46	45	45.5
26	4.9	9.4	7.15	16.0	23.5	19.75	42	49	45.5
27	0.9	8.5	4.7	16.7	21.7	19.2	64	51	57.5
28	2.3	8.6	5.45	14.2	23.2	18.7	46	46	46.0
29	4.2	9.8	7.0	15.6	23.4	19.5	42	49	45.5
30	10.2	11.5	10.85	18.4	23.0	20.7	44	50	47.0
31	5.5	11.9	8.7	18.3	23.4	20.85	68	58	63.0
Average	**4.01**	**7.50**	**5.76**	**14.72**	**20.15**	**17.44**	**55.09**	**51.25**	**53.18**

Table 11.6: (June)

	Minimum (°C)			Maximum (°C)			Humidity (%)		
Date	**2005**	**2006**	**Average**	**2005**	**2006**	**Average**	**2005**	**2006**	**Average**
1	3.1	9.5	6.3	13.4	21.6	17.5	63	66	64.5
2	3.4	10.7	7.05	16.1	17.9	17.0	43	70	56.5
3	4.8	10.4	7.6	17.9	20.7	19.3	50	52	51.0
4	6.0	8.4	7.2	18.5	20.0	19.25	54	64	59.0
5	6.7	9.9	8.3	20.0	19.5	19.75	62	57	59.5
6	7.6	5.2	6.4	18.9	18.8	18.85	74	65	69.5
7	8.4	8.7	8.55	15.2	22.5	18.85	56	56	56.0
8	5.6	9.9	7.75	15.3	21.9	18.6	60	57	58.5
9	6.2	8.5	7.35	13.5	23.5	18.5	68	52	60.0
10	3.4	11.5	7.45	13.6	23.8	18.7	64	54	59.0
11	4.7	9.7	7.2	16.2	21.6	18.9	63	52	57.5
12	5.4	6.3	5.85	19.4	18.4	18.9	45	64	54.5
13	8.1	5.6	6.85	19.3	15.1	17.2	60	56	58.0
14	6.8	6.5	6.65	18.9	20.1	19.5	67	72	69.5
15	8.6	10.7	9.65	20.1	18.4	19.25	59	58	58.5
16	8.4	6.5	7.45	18.9	18.4	18.65	56	79	67.5
17	6.7	4.9	5.8	17.6	10.3	13.95	61	67	64.0
18	8.9	4.6	6.75	21.6	11.4	16.5	57	49	53.0
19	11.5	3.7	7.6	21.0	15.4	18.2	58	45	51.5
20	11.7	10.3	11.0	23.4	18.3	20.85	54	41	47.5
21	10.5	7.2	8.85	22.3	20.0	21.15	51	42	46.5
22	8.9	6.7	7.8	22.4	21.8	22.1	49	41	45.0
23	11.1	10.4	10.75	23.6	24.4	24.0	44	46	45.0
24	13.3	13.6	13.45	24.9	26.2	25.55	32	50	41.0
25	9.8	13.8	11.8	26.0	27.0	26.5	39	42	40.5
26	12.7	14.2	13.45	28.5	28.0	28.25	41	47	44.0
27	13.3	18.4	15.85	27.9	27.6	27.75	43	50	46.5
28	14.5	11.5	13.0	28.3	24.7	26.5	44	64	54.0
29	14.7	15.1	14.9	27.8	25.5	26.65	43	51	47.0
30	14.8	12.1	13.45	28.4	24.8	26.6	48	60	54.0
Average	**8.65**	**9.47**	**9.06**	**20.63**	**20.92**	**20.77**	**53.6**	**55.63**	**54.62**

Table 11.7: (July)

	Minimum (°C)			Maximum (°C)			Humidity (%)		
Date	2005	2006	Average	2005	2006	Average	2005	2006	Average
1	14.3	10.4	12.35	26.6	24.9	25.75	48	61	54.5
2	16.5	10.8	13.65	26.6	24.6	25.6	48	55	51.5
3	11.8	11.3	11.55	23.1	25.1	24.1	55	50	52.5
4	11.1	16.7	13.9	22.4	28.0	25.2	52	48	50.0
5	14.8	17.2	16.0	23.8	27.0	25.4	54	48	51.0
6	10.9	16.0	13.45	15.9	26.5	21.2	65	47	56.0
7	7.6	15.1	11.35	15.1	29.2	22.15	79	46	62.5
8	9.3	17.6	13.45	12.9	32.8	22.85	67	47	57.0
9	11.8	17.3	14.55	21.1	33.1	27.1	61	45	53.0
10	12.6	20.0	16.3	26.5	33.0	29.75	52	49	50.5
11	14.4	20.1	17.25	28.4	31.8	30.1	52	51	51.5
12	18.0	16.6	17.3	29.3	29.7	29.5	56	55	55.5
13	13.0	14.7	13.85	24.2	22.3	23.25	74	53	63.5
14	13.7	6.4	10.05	20.9	22.7	21.8	69	79	74.0
15	13.5	8.8	11.15	22.4	13.0	17.7	67	66	66.5
16	17.4	11.4	14.4	27.2	23.5	25.35	50	53	51.5
17	14.8	11.4	13.1	26.3	26.6	26.45	49	50	49.5
18	14.2	14.3	14.25	25.2	27.4	26.3	51	50	50.5
19	11.7	17.3	14.5	24.8	29.3	27.05	41	52	46.5
20	11.8	18.9	15.35	25.9	29.5	27.7	44	52	48.0
21	13.9	15.3	14.6	28.3	28.3	28.3	44	54	49.0
22	16.9	20.0	18.45	27.9	28.5	28.2	48	51	49.5
23	13.8	16.7	15.25	26.8	31.1	28.95	50	49	49.5
24	15.2	15.5	15.35	27.3	33.1	30.2	50	65	57.5
25	14.2	17.7	15.95	24.8	29.2	27.0	52	55	53.5
26	17.6	20.1	18.85	27.6	30.6	29.1	49	56	52.5
27	18.7	18.0	18.35	26.4	29.0	27.7	49	56	52.5
28	17.7	14.5	16.1	27.9	29.8	28.85	52	56	54.0
29	17.4	16.0	16.7	28.5	29.6	29.05	51	62	56.5
30	18.0	14.0	16.0	28.1	26.8	27.45	49	66	57.5
31	15.4	11.5	13.45	29.2	31.7	30.45	52	70	61.0
Average	**14.25**	**15.21**	**14.74**	**24.88**	**26.44**	**25.97**	**54.19**	**54.74**	**54.46**

Table 11.8: (August)

	Minimum (°C)			Maximum (°C)			Humidity (%)		
Date	2005	2006	Average	2005	2006	Average	2005	2006	Average
1	14.7	11.4	13.05	29.4	31.6	30.5	53	70	61.5
2	17.0	11.9	14.45	28.3	25.4	26.85	52	78	65.0
3	17.4	12.9	15.15	29.9	23.8	26.85	51	82	66.5
4	17.0	13.4	15.2	30.4	22.2	26.3	49	73	61.0
5	16.2	13.7	14.95	29.4	21.8	25.6	45	77	61.0
6	13.8	14.1	13.95	28.5	18.8	23.65	51	70	60.5
7	11.7	12.9	12.3	22.0	22.0	22.0	51	73	62.0
8	11.7	12.6	12.15	24.3	19.1	21.7	67	71	69.0
9	10.5	11.8	11.15	28.8	19.2	24.0	70	75	72.5
10	13.3	12.9	13.1	23.7	24.8	24.25	51	56	53.5
11	11.7	12.9	12.3	25.2	25.7	25.45	50	81	65.5
12	13.4	13.3	13.35	26.8	21.0	23.9	52	68	60.0
13	14.9	14.9	14.9	27.8	23.9	25.85	53	59	56.0
14	13.1	14.4	13.75	28.3	24.2	26.25	50	57	53.5
15	12.7	12.8	12.75	29.2	23.9	26.55	52	56	54.0
16	18.1	13.2	15.65	31.1	23.4	27.25	45	62	53.5
17	16.7	16.4	16.55	28.5	25.6	27.05	50	49	49.5
18	11.2	15.8	13.5	21.1	24.6	22.85	51	48	49.5
19	12.9	12.7	12.8	25.9	26.2	26.05	51	51	51.0
20	13.3	13.2	13.25	26.2	28.4	27.3	46	58	52.0
21	16.3	14.5	15.4	25.6	29.1	27.35	49	66	57.5
22	12.7	14.4	13.55	26.5	25.5	26.0	52	58	55.0
23	13.8	13.2	13.5	26.9	26.0	26.45	51	57	54.0
24	13.3	15.1	14.2	26.8	27.1	26.95	49	54	51.5
25	11.3	18.0	14.65	26.2	28.3	27.25	52	51	51.5
26	9.4	15.0	12.2	25.3	27.0	26.15	50	49	49.5
27	9.9	13.8	11.85	24.2	26.8	25.5	47	48	47.5
28	9.0	11.4	10.2	24.0	26.2	25.1	51	82	66.5
29	8.6	9.4	9.0	21.0	15.9	18.45	53	70	61.5
30	7.2	8.5	7.85	22.1	19.0	20.55	49	70	59.5
31	8.2	8.5	8.35	23.7	20.4	22.05	50	56	53.0
Average	**12.93**	**13.20**	**13.06**	**26.35**	**24.11**	**25.22**	**51.38**	**63.70**	**57.55**

Table 11.9: (September)

	Minimum (°C)			Maximum (°C)			Humidity (%)		
Date	2005	2006	Average	2005	2006	Average	2005	2006	Average
1	8.3	8.5	8.4	26.0	23.1	24.55	43	70	56.5
2	9.3	8.6	8.95	27.5	14.8	21.15	43	78	60.5
3	9.9	8.4	9.15	27.3	11.9	19.6	47	75	61.0
4	10.7	7.3	9.0	26.4	19.0	22.7	47	70	58.5
5	10.4	7.6	9.0	27.9	20.8	24.35	44	60	52.0
6	10.7	9.7	10.2	28.8	22.4	25.6	43	57	50.0
7	11.8	11.7	11.75	29.0	23.7	26.35	42	55	48.5
8	12.3	11.6	11.95	29.3	24.8	27.05	54	57	55.5
9	16.2	10.0	13.1	27.2	25.1	26.15	55	55	55.0
10	15.8	9.2	12.5	23.6	25.2	24.4	53	54	53.5
11	10.1	8.9	9.5	23.3	25.7	24.5	56	52	54.0
12	12.5	12.4	12.45	18.1	23.0	20.55	59	51	55.0
13	11.9	8.5	10.2	21.6	21.1	21.35	58	58	58.0
14	4.1	11.5	7.8	13.5	22.7	18.1	75	57	66.0
15	8.5	12.6	10.55	17.0	23.1	20.05	58	50	54.0
16	8.0	11.6	9.8	21.9	21.8	21.85	54	54	54.0
17	9.3	6.8	8.05	24.1	21.7	22.9	52	50	51.0
18	14.3	6.5	10.4	24.8	23.1	23.95	55	48	51.5
19	4.2	11.7	7.95	17.5	24.6	21.05	70	48	59.0
20	5.8	11.7	8.75	20.2	23.8	22.0	65	50	57.5
21	5.9	11.7	8.8	21.1	20.8	20.95	48	50	49.0
22	6.2	8.3	7.25	22.0	20.8	21.4	47	58	52.5
23	8.7	4.3	6.5	20.7	20.8	20.75	55	58	56.5
24	3.8	4.8	4.3	15.3	20.8	18.05	70	52	61.0
25	6.8	4.2	5.5	20.5	20.8	20.65	50	47	48.5
26	6.9	3.7	5.3	21.2	20.8	21.0	51	56	53.5
27	7.1	3.6	5.35	20.3	20.8	20.55	54	53	53.5
28	4.6	2.6	3.6	19.7	20.8	20.25	50	50	50.0
29	7.3	2.8	5.05	18.6	20.8	19.7	48	51	49.5
30	5.2	3.8	4.5	18.6	19.1	18.85	48	50	49.0
Average	**8.88**	**8.15**	**8.52**	**22.43**	**21.59**	**22.01**	**53.13**	**55.8**	**54.46**

Table 11.10: (October)

	Minimum (°C)			Maximum (°C)			Humidity (%)		
Date	2005	2006	Average	2005	2006	Average	2005	2006	Average
1	4.7	3.5	4.1	17.6	19.1	18.35	50	50	50.0
2	2.2	3.6	2.9	19.1	19.6	19.35	49	49	49.0
3	3.7	8.6	6.15	19.7	19.1	19.4	48	50	49.0
4	3.7	3.6	3.65	19.8	17.1	18.45	49	54	51.5
5	2.7	1.0	1.85	18.7	15.8	17.25	52	55	53.5
6	3.7	3.3	3.5	19.2	18.4	18.8	49	47	48.0
7	3.6	1.6	2.6	18.9	17.9	18.4	48	50	49.0
8	2.4	0.9	1.65	19.1	17.7	18.4	45	48	46.5
9	7.7	2.5	5.1	17.5	20.1	18.8	51	49	50.0
10	3.1	3.3	3.2	15.6	19.0	17.3	57	50	53.5
11	6.1	3.4	4.75	16.1	18.2	17.15	56	58	57.0
12	3.7	3.3	3.5	12.3	17.7	15.0	60	54	57.0
13	-1.4	5.4	2.0	9.7	17.7	13.7	70	59	64.5
14	0.4	4.0	2.2	13.0	13.8	13.4	58	62	60.0
15	-0.9	1.0	0.05	11.9	12.4	12.15	55	68	61.5
16	-0.7	1.0	0.15	14.6	13.4	14.0	48	60	54.0
17	-2.5	0.9	-0.8	12.2	13.2	12.7	60	57	58.5
18	-1.0	-0.4	-0.7	11.9	14.8	13.35	51	54	52.5
19	-3.0	2.8	-0.1	10.3	14.7	12.5	49	57	53.0
20	-5.5	2.4	-1.55	9.8	11.7	10.75	49	60	54.5
21	-4.4	2.0	-1.2	12.2	12.5	12.35	46	80	63.0
22	-4.6	-0.3	-2.45	13.1	6.3	9.7	45	70	57.5
23	-4.8	-2.0	-3.4	14.9	8.3	11.6	44	70	57.0
24	-3.8	-2.5	-3.15	13.6	11.0	12.3	46	58	52.0
25	-1.0	-2.0	-1.5	15.3	10.8	13.05	49	57	53.0
26	-1.5	-0.5	-1.0	12.2	13.8	13.0	58	58	58.0
27	-4.4	1.2	-1.6	10.1	11.6	10.85	53	80	66.5
28	-4.9	-1.0	-2.95	11.1	8.0	9.55	50	55	52.5
29	-3.6	-3.5	-3.55	12.1	9.8	10.95	49	54	51.5
30	-5.6	-3.9	-4.75	11.3	10.7	11.0	56	55	55.5
31	-4.7	-3.0	-3.85	10.7	12.6	11.65	47	50	48.5
Average	**-0.34**	**1.29**	**0.47**	**14.40**	**14.41**	**14.36**	**51.51**	**57.25**	**54.42**

Table 11.11: (November)

	Minimum (°C)			Maximum (°C)			Humidity (%)		
Date	**2005**	**2006**	**Average**	**2005**	**2006**	**Average**	**2005**	**2006**	**Average**
1	-3.9	-3.5	-3.7	10.0	12.5	11.25	50	51	50.5
2	-2.7	-4.2	-3.45	7.3	13.0	10.15	70	48	59.0
3	-4.6	-2.4	-3.5	5.7	12.2	8.95	72	53	62.5
4	-8.2	-3.6	-5.9	5.4	11.9	8.65	56	55	55.5
5	-7.5	-4.3	-5.9	6.2	12.0	9.1	57	45	51.0
6	-8.3	-4.3	-6.3	5.2	13.0	9.1	59	51	55.0
7	-8.8	-2.6	-5.7	6.2	13.8	10	58	53	55.5
8	-3.9	-4.4	-4.15	6.8	11.8	9.3	53	56	54.5
9	-7.2	-5.0	-6.1	6.1	12.0	9.05	65	54	59.5
10	-7.6	-3.9	-5.75	8.1	12.0	10.05	60	51	55.5
11	-3.5	-3.9	-3.7	11.2	12.0	11.6	54	55	54.5
12	-5.0	-3.9	-4.45	8.3	11.2	9.75	66	51	58.5
13	-7.8	-3.9	-5.85	8.2	12.0	10.1	60	60	60.0
14	-9.5	-3.9	-6.7	6.3	12.0	9.15	59	51	55.0
15	-7.2	-6.3	-6.75	6.2	12.0	9.1	50	55	52.5
16	-9.1	-6.3	-7.7	10.2	11.2	10.7	51	61	56.0
17	-9.8	-6.3	-8.05	6.2	11.2	8.7	52	66	59.0
18	-12.4	-6.3	-9.35	4.7	11.2	7.95	54	67	60.5
19	-10.8	-6.3	-8.55	6.0	11.2	8.6	51	72	61.5
20	-8.0	-7.7	-7.85	7.1	11.2	9.15	50	66	58.0
21	-4.9	-7.3	-6.1	8.0	11.2	9.6	55	62	58.5
22	-3.3	-3.4	-3.35	7.3	8.0	7.65	75	82	78.5
23	-5.0	-5.8	-5.4	6.3	5.0	5.65	70	72	71.0
24	-9.9	-9.0	-9.45	5.8	3.1	4.45	65	68	66.5
25	-11.1	-11.0	-11.05	6.0	2.5	4.25	63	62	62.5
26	-9.0	-10.3	-9.65	6.6	1.7	4.15	54	60	57.0
27	-7.4	-6.8	-7.1	3.5	4.0	3.75	60	51	55.5
28	-7.7	-8.4	-8.05	3.7	0.3	2.0	73	71	72.0
29	-8.8	-7.3	-8.05	3.5	2.7	3.1	61	63	62.0
30	-12.1	-12.8	-12.45	5.7	2.4	4.05	59	66	62.5
Average	**-7.5**	**-5.83**	**-6.69**	**6.59**	**9.34**	**7.97**	**59.4**	**57.6**	**59.33**

Table 11.12: (December)

Date	Minimum (°C)			Maximum (°C)			Humidity (%)		
	2005	2006	Average	2005	2006	Average	2005	2006	Average
1	-12.6	-12.7	-12.65	5.7	4.6	5.15	59	61	60.0
2	-11.2	-11.0	-11.1	3.7	3.9	3.8	58	56	57.0
3	-11.4	-6.6	-9.0	5.1	2.5	3.8	56	58	57.0
4	-10.4	-3.1	-6.75	0.6	2.2	1.4	72	86	79.0
5	-11.3	-2.2	-6.75	1.8	1.3	1.55	72	78	75.0
6	-10.2	-7.1	-8.65	2.5	5.5	4.0	64	73	68.5
7	-12.6	-9.5	-11.05	5.9	6.7	6.3	65	73	69.0
8	-13.4	-10.0	-11.7	5.1	3.7	4.4	63	72	67.5
9	-11.8	-13.0	-12.4	4.0	0.4	2.2	59	75	67.0
10	-11.8	-10.8	-11.3	3.6	-0.3	1.65	72	84	78.0
11	-14.9	-8.7	-11.8	0.8	-2.4	-0.8	67	86	76.5
12	-15.8	-12.3	-14.05	2.0	-3.3	-0.65	60	85	72.5
13	-16.6	-14.0	-15.3	3.0	-2.4	0.3	60	75	67.5
14	-16.0	-16.0	-16.0	3.0	-1.5	0.75	58	75	66.5
15	-15.9	-14.2	-15.05	3.0	-0.5	1.25	58	75	66.5
16	-15.2	-13.5	-14.35	1.0	2.1	1.55	59	71	65.0
17	-12.5	-10.2	-11.35	0.4	5.0	2.7	60	70	65.0
18	-12.3	-14.0	-13.15	1.1	4.0	2.55	70	74	72.0
19	-14.4	-15.0	-14.7	2.5	1.9	2.2	68	75	71.5
20	-14.0	-14.1	-14.05	4.0	1.2	2.6	65	72	68.5
21	-10.4	-12.5	-11.45	0.7	-2.5	-0.9	61	87	74.0
22	-9.0	-12.5	-10.75	-1.3	-3.1	-2.2	85	82	83.5
23	-13.0	-15.2	-14.1	-1.2	1.9	0.35	74	75	74.5
24	-15.6	-14.0	-14.8	-1.6	1.2	-0.2	70	65	67.5
25	-15.8	-12.0	-13.9	-0.1	2.0	0.95	51	67	59.0
26	-14.0	-10.0	-12.0	1.6	-0.5	0.55	60	62	61.0
27	-10.9	-9.0	-9.95	1.9	2.1	2.0	59	80	69.5
28	-8.9	-14.9	-11.9	1.2	1.5	1.35	85	75	80.0
29	-12.6	-15.0	-13.8	1.1	0.7	0.9	60	60	60.0
30	-13.1	-11.2	-12.15	1.9	2.8	2.35	61	69	65.0
31	-10.2	-13.2	-11.7	2.6	5.5	4.05	47	61	54.0
Average	**-12.83**	**-11.53**	**-12.18**	**5.34**	**1.44**	**1.80**	**63.80**	**72.8**	**68.30**

References

1. Anonymous: Ware and seed potato production technique in North-West hills. Extension Bulletin No. 15. CPRI, Shimla.

2. Anonymous, 1991. Manual for True Potato Seeds (TPS) production and utilization. Extension Bulletin No. 30 (E) ,CPRI ,Shimla.

3. Anonymous, 1993. Potato storage and processing. Advances in Horticulture Vol. 7 – Potato. Eds: K.L. Chadha and J.S. Grewal, Malhotra Publishing House, New Delhi. 703–715.

4. Anonymous, 2002. CIPC suppliers prepared for residue limit. *Potato Review* (November): 20–23.

5. Ahire, N.R., P.G. Bhoi and A,V, Solanke. 2000. Effect of row spacing and planting system on growth and yield of potato under surface and drip irrigation. *J. Indian Potato Assoc.* 27 (1–2): 59–60.

6. Arteca, R.N. 1982. Effect of root applications of kinetin and gibberellic acid on transplanting shock in tomato plants. *Hort Science.* 17: 633–634.

7. Aulakh, P.S. and H.S. Sur. 1999. Effect of mulching on soil temperature, soil moisture, weed population, growth and yield in pomegranate. *Prog. Hort.* 31 : 131–133.

8. Awashti, R.P. and M. Singh. 1982. Growth and yield of potato influenced by seed size and spacing in Khasi Hills of Meghalaya. In *Potato in Developing Countries,* B.B. Nagaich, et al, (ed), Indian Potato Association: 198–204.

9. Azariah, M.D. and K. Saptharishi. (1951). Influence of seed size and spacing on the yield of potatoes. *Madras Agriculture Journal* 38: 503–507.

10. Bajwa, M.S., S.S. Deol. and J.C. Bakshi. 1970. Effect of mulching on soil moisture, soil temperature, weed density, yield and fruit quality in grapes. *Punjab Hort. J.* 10: 232–239.

11. Bharat, B. and C.L. Acharya. 2000. Root water uptake as influenced by mulched and unmulched treatments in potato. *J. of Indian Potato Assoc.* 27 : 41–44.

12. Bhardwaj, V., S.K. Kaushik, S.K. Chakrabarti, S.K. Pandey, P.H. Singh, P. Manivel and B.P. Singh. 2007. Combining resistance to late blight and PVY in Potato. *Potato J.* 34 (1–2): 41–42.

13. Bhattacharjee, A.K. and V.S. Khuswah. 1988. Feasibility of minimum tillage and cultural practices in potato (*Solanum tuberosum* L.) cultivation. *Indian Journal of Agriculture Sciencies* 58: 267–273.

14. Bhutani, R.D. and S.C. Khurana. 2005. Storage behavior of potato genotypes under ambient conditions. *Potato Journal*. 32 (3-4): 209–210.

15. Bremner, P.M. and M.A. Taha. 1966. Studies in potato agronomy.I. The effect of variety, seed size and spacing on growth, development and yield. *J. Agric. Camb.* 66: 241–252.

16. Bremner, P.M. and R.W. Radley. 1966. Studies in potato agronomy. II. The effect of variety, seed size and spacing on growth, development and yield. *J. Agric. Camb.* 66: 253–262.

17. Chandla, V.K., B. Paul, A. Basu and M. Chettri. 2003. Management of potato tuber moth in country stores with CIPC: A possibility under testing. *J. Indian Potato Assoc.* 30 (1–2):153–154.

18. Chaudhuri, S. Sen., S. Pal and M.D. Upadhya, 1990. TPS for tuber production in Tripura. In: Commercial Adoption of TPS Technology: Prospects and Problems [P.C. Gaur ed.]. CPRI, Shimla: 85–89.

19. Chaurasia, O.P. and B. Singh. 1996. Cold desert plants–Vol–1, Leh Valley – Field Research Laboratory(FRL) DRDO, Leh (J &K).

20. Dabas, T.P.S., Hira Nand and M.S. Kadian. 1993. Performance of true potato seed under lab to land on farm demonstration in Meerut. *J. Indian Potato Assoc.* 20:30.

21. Deka, N.C., Z. Ullah and T.C. Dutta. 1997. Yield performance and economics of potato cultures under late planting conditions. *J. Indian Potato Assoc.* 24 (1-2): 68–70.

22. Dhatt, A.S., B.V.C. Mahajan, K.S. Sandhu, A. Garg and Satish Sharma. 2003. Bio efficacy studies of Chlorpropham on stored potatoes. *J. Indian Potato Assoc.* 30 (1–2): 173–174.

23. Dubey, A.K. 2003. Studies on planting geometry and plant population of TPS. *J. Indian Potato Assoc.* 30 (1–2): 27–28.

24. Dutt, B.L. 1979. Bacterial and fungal diseases of potato. Tech. Bulletin, ICAR, New Delhi: 199.

25. Environmental Protection Agency. 1996. Registration eligibility decision. EPA–738–R–96–023.

26. Esu, K. 1977. Anatomy of Seed Plants. 2nd ed. Wiley, New York.

27. Ezekiel, R. 1997. Effect of environmental and cultural factors during growth of seed potato (*Solanum tubersum*) performance of progeny tuber as seed. *Indian Journal Agriculture Sciences.* 67 (8) : 308-311.

28. Ezekiel, R., B. Singh., D. Kumar., V. Paul and M. Das. 2005. Storage of potatoes at 10–12^{0}C with CIPC treatment in commercial cold store. *Potato Journal.* 32 (3-4): 211–212.

29. Ezekiel, R., B. Singh and Sumi Khurana. 2003. Potato recipes Technical Bulletin No. 55 (Revised). CPRI, Shimla.

30. Garg I.D., V.K. Chandla, A.M. Paul Khurana and N. Singh. 2003. Potato cultivation and health status of the crop in Leh, Ladakh (J & K). *J. Indian Potato Assoc.* 30 (1- 2): 121–122.

31. Gaur, P.C., P.S. Naik, S.K. Kaushik and S.K. Chakrabati. 1999. Indian potato varieties. CPRI, Shimla: 51.

32. Gaur, P.C. and S.K. Pandey. 1990. Adoption of TPS technology in India. In: commercial adoption of true potato seed technology-prospects and problems (PC Gaur, Rd). CPRI, Shimla: 7–10.

33. Gomez, K.A. and A.A. Gomez, 1984. Statistical procedure for Agricultural Research. John Wiley and Sons, Singapore.

34. Gopal, J. 1994. Flowering behavior made sterility and berry setting in tetraploid *Solanum tuberosum* germplasm. *Euphytica.* 72: 133–142.

35. Gopal, J. 2001. Genetic parameters and character associates for family selection in potato breeding programmes. *Journal Genetics & Breeding* 55: 201–208.

36. Gopal, J. 2003. True Potato seed: Breeding for hardiness. *Journal of New Seeds.* 5 (4). 57–73.

37. Grewal, J.S., R.C. Sharma and S.S. Saini. 1992. Agro techniques for intensivation in India. Indian Council of Agricultural Research Krishi Anusandhan Bhavan, Pusa, New Delhi-110012.

38. Grewal, J.S. and S.N. Singh. 1980. Effect plants and yield in alluvial soils of Punjab. *Plant and soil.* 57: 105–110.

39. Guarda and S. Giulliari. 1983. Effect of plant density on seed tuber yield on potato varieties.*Sementi-Ellette*. 29: 23 – 28 (Plant Breeding Abst. 54 : 3237).

40. Gupta, V.K., S. Kumar, U. Sah, S.K. Pandey and M.B. Gogoi. 2003. Assessment of seedling tubers as seed under rainfed condition of Meghalaya. *J. Indian Potato Assoc*. 30 (1–2): 51–52.

41. Iritani, W.M, R. Thorton, L. Welter and O. Leavy. 1972. Relationship of seed size, spacing, stem numbers to yield of Russet Burbank potatoes. *Am J. Potato Reserch* 40: 463–479.

42. Kabir, M.H. 1991. Studies on the population density of a hybrid TPS progeny. *Haryana J. Horti Sci*. 20: 125–128.

43. Kadian. M.S., P.K. Patel, M.D. Upadhya and K.C. Thakur. 1987. Evaluation of TPS progenies for the potato production in Deesa (Gujarat) *J. Indian Potato Assoc*. 14 : 57–59.

44. Kadian, M.S., P.K. Patel. K.C., Thakur and M.D. Upadhya. 1988. Comparative yield potential of seedling tuberlets from true potato seed in Deesa (Gujrat). *J. Indian Potato Assoc*.15: 68–73.

45. Kadian, M.S., Sarath G Ilangantileke, Mpoeen UD Din, Nasrat Waşsimi and Lal Mohammad. 2003. CIP in Afghanistan: Production of good quality potato seed to meet the emergency seed requirement. *J. of Indian Potato Assoc* 30: 55–56.

46. Kadian, M.S., S.G.Ilangantileke, M. Hossain, A. Mahmud and C. Crissman. 2003. Adoption of improved technology for sustained production of good quality potato seed in the informal system of Bangladesh. 6(2): 1–5.

47. Kaul, H.N. and A. Mehta. 1988. Dormancy periods in commercial Indian potato varieties. *Proc.Indian National Academy Sci*. 58 (B) II. 285–290.

48. Kaul, H.N. and A. Mehta. 1999. Storage of potatoes in India. CPRI, Shimla. 47.

49. Kaul, H.N. and N.P. Sukumaran. 1984. Effect of late harvesting and delayed dispatch to cold storage on seed potatoes. *J. Agric. Sci* 46: 338–342.

50. Khan, I.A., A.H. Gillani and M. Maqbool. 1980. Effect of chilling and seed tuber size on the growth and yield of potato varieties. *J. Agric. Res. Pakistan* 18: 103–116.

51. Khurana, S.C., M.L. Pandita and V.K. Srivastwa. 1991. Effect of seed size and seed rate on potato yield. *J. Indian Potato Assoc.* 18: 167–168.

52. Kleinkopf, G.E., T.L. Brandt, M.J. Frazier and G. Moller. 1997. CIPC residues on stored russet Burbank potatoes: 1 Maximum label application. *American Potato Journal*. 74: 107–117.

53. Kumar, D., H.N. Kaul and S.V. Singh. 1995. Keeping quality in advance potato selection during non – refrigerated storage. *J. Indian Potato Assoc.* 22: 105–108.

54. Kumar D., P.C. Gaur, S.K. Pandey, J. Singh and S. V. Singh. 1999. Effect of storage and reconditioning on recently released Indian potato processing varieties. J. *Indian Potato Assoc.* 26 (1& 2): 71–73.

55. Kumar, D and R. Ezekiel. 2002. Sugar accumulation in potato tubers at two different temperatures of storage. Proceedings International Conference on vegetables. 413–415.

56. Kumar, J., J.L. Mangal, R.D. Bhutani and S.C. Khurana. 2001. Evaluation of shelf life of potato hybrids under ambient conditions. *J. Indian Potato Assoc.* 28: 283–284.

57. Kumar, N., S. Chandra and A.K. Shrivastva. 2006. Mulching: An effective tool for sustainable agriculture in rainfed farming. *Indian Farmers' Digest*. June. 30–33.

58. Kumar, R and B.K. Shrivastava. 1998. Effect of different plastic and organic mulches on growth characteristics of winter tomato. *Agri. Sci. Digest*. 18 : 123–125.

59. Kumar, R and Jai Gopal. 2003. Combining ability of Andigena accessions for yield components and tuber dry matter in third clonal generation. *J. Indian Potato Assoc.* 30 (1 – 2): 3–4.

60. Kumar, S.V., S.K. Pandey and S.M. Paul Khurana. 2001. Storage behavior of some advanced hybrids in plain of western UP at ambient temperature. *J. Indian Potato Assoc.* 28: 135–136.

61. Kumar, S., R. Ezekiel ans B. Singh. 2003. Weight loss of tubers and chip colour of potato cultivars stored at different temperatures after CIPC treatment. *J. Indian Potato Assoc.*30 (1 – 2): 175–176.

62. Khurana, S.C., Srivastwa, V.K. and Pandita, M.L. 1989. Package of practices for table and seed potato production in Haryana. Haryana Agriculture University, Hisar: 9.

63. Khurana, S.M. Paul. 1999. Potato viruses and viral diseases. Technical Bulletin No. 35 (revised). CPRI, Shimla: 94.

64. Lal, S.S., T.S. Sahota and J.S. Grewal. 1981. Studies of seed size and spacing on potato (*Solanum tubersum* L) for optimum tuber yield. *J. Indian Potato Assoc.* 8: 74–80.

65. Lee, C.L. and W.P. Hackett. 1976. Root regeneration of transplanted Pistacia chinensis Bunge seedlings at different growth stages. *Journal American Society Horticulture Science* 101: 236–240.

66. Luthra, S.K., P.C. Sharma and J. Gopal. 2001. Interrelationship between genetic parameters for tuber yield in potato. *J. Indian Potato Assoc.* 28: 15–17.

67. Luthra, S.K., P.C. Pandey, S.V. Singh, S.K. Pandey, S.M. Khurana, I.A. Khan and B.P. Singh. 2006. Kufri Arun: A new red skin potato variety. *Potato J.* 33 (1–2): 20–25.

67. Madalageri, M.B., V.M. Ganiger, R.M. Hosamani and B.B. Madalageri. 1997. True potato seed quality as influenced by genotypes. *J. Indian Potato Assoc.* 24 (1–2): 82–83.

68. Mahajan, B.V.C., K.S. Sandhu and A.S. Dhatt. 2003. High temperature storage of potato with CIPC treatment. In: Abstract, Conference on processing and export potential of potatoes within Asia, March 10, 2003, IPA, CPRI and CIP, Modipuram, Meerut:38–39.

69. Mahmood Ahmed and I.A. Gill. 1984. Relationship of stem density with seed size weight in four potato cultivars (*Solanum tubersum* L). *J. Agric Res. Pakistan* 22: 309–312.

70. Mallet, J.B., R.J. Rutherford and R. D. Mouuld. 1973. Many stems – high yields. *Farming S. Africa.* 48: 8–10.

71. Mangal, J.L., J. Kumar, R.D. Bhutani and S.C. Khurana. 2002. Comparative shelf life of seven potato cultivars in evaporative cool chambers and ambient conditions. *Haryana J. Hort. Sci.* 31: 272–74.

72. Martin, M.H. 1983. Techniques for successful field seedlings of TPS. *Amer. Potato J.* 60: 245–59.

73. Marwaha R.S., R. Ezekiel and B.P. Singh. 2003. Post – harvest and pre – holding changes in sugar content of potatoes. *J. Indian Potato Assoc.*30 (1–2): 191–192.

74. Mathur, P.B. 1951. Studies in cold storage of potatoes. *J.Sci. Industr. Res.*: 224–228.

75. Mathur, P.B., K. Kripal Singh and N.S. Kapur. 1952. Cold storage studies in certain potato varieties. *J. Indian Sco. Refrig. Engineers.* 3: 31–39.

76. Mathur, P.B., K. Kripal Singh and N.S. Kapur. 1952. Loss in weight, sprouting and change in sugar content in potato tubers stored at various temperatures. *J. Indian Sco. Refrig. Engineers.* 3: 40–49.

77. Mehta, A., and H.N. Kaul. 1987. Storage behavior of potato cultivars in evaporatively cooled store. *J. Indian Potato Assoc.* 14(1&2): 69–71.

78. Mehta, A., and H.N. Kaul. 1988. High temperature storage of potato (*Solanum tuberosum*) for processing–1 feasibility study. *Plant Fd. Human Nutr.* 38: 263–268.

79. Mehta, A., and H.N. Kaul. 1990. Contribution of transpiration and respiration towards weight loss in potatoes during non-refrigerated storage. Souvenir and Abstract. National symposium on strategies for production, marketing, storage and processing, New Delhi, India: 81.

80. Mehta, A. 2005. Commercial potential of CIPC for sprout inhibition in potatoes under heap storage. *Potato Journal.* 32 (3- 4). 203–204.

81. Mehta A., B. Singh., D. Kumar and R. Ezekiel. 2007. Evaluation of CIPC spray for sprout inhibition in potatoes under traditional storage methods. *Potato Journal.* 34 (1–2): 67–68.

82. Midmore, D.J. 1984. Potato (*Solanum* spp.) in the hot tropics. I. Soil temperature effect on emergence plant development and yield. *Field crops Res.* 8 : 255–71.

83. Midmore, D.J. 1988. Potato in hot tropics. VI plant population effect on soil moisture, plant development and tuber yield. *Fld. Crop Res.* 19: 183–200.

84. Minhas, J.S. 1989. A simple country store for table potatoes. *J. Indian Potato Assoc.* 16: 125–126.

85. Mishra, A.C., N.P. Singh and H.S. Ram. 2003. Genetic divergence among advanced hybrids and varieties of potato. *J. Indian Potato Assoc.* 29: 175–176.

86. Moolani, M.K. and S.B. Hukkeri. 1963. Effect of cultivation depth, fertilizers and irrigation on the yield of potato. *J.S.D. and Engg. Res.* 7: 368–373.

87. Nandekar, D.N. 2005. Effect of seedling tuber size and fertilizer levels on growth, yield and economics of potato production. *Potato J.* 32 (1–2): 71–73.

88. Pandey, P.C., B.H. Girish, M.S. Kadian and S.G. Ilngantileke. 2003. Evaluation of true potato seed families for adaptability and yield characteristics. *J. Indian Potato Assoc.* 30 (1 – 2): 25–26.

89. Pandey, P.C., M.S. Kadian and M.D. Upadhya. 1990. Field performance of TPS tubers in three successive clonal generations. *J. Indian Potato Assoc.* 17: 30–33.

90. Pandey, P.C., S.K. Luthra, B.P. Singh and S.K. Pandey. 2005. Selection of superior crosses on the basis of progeny mean in potato. *Potato J.* 32 (3–4): 127–128.

91. Pandey, S.K. and P.K. Gupta. 1995. Comparision of Andigena and Tuberosum for enhancing TPS production. *J. Indian Potato Assoc.* 22: 122–128.

92. Pandey, S.K., R.S. Marwaha, P. Manivel and R. Kumar. 2005. Evaluation of potato germplasm for cold chipping ability and cold sweetening. *Potato Journal*. 32 (3-4). 121–122.

93. Patel, C.K., S.H. Patel, G.V. Sasani, R.N. Patel and N.H. Patel. 2003. Response of potato Cv. Kufri Badshah to inter and intra row spacing under Gujarat Condition. *J. Indian Potato Assoc.* 30 (1–2): 49–50.

94. Patel, H.R., A.M. Shekh, G.C. Patel, D.S. Mistry and C.T. Patel. 1999. Influence of soil temperature on potato emergence. *J. of Indian Potato Assoc.* 26 : 23–26.

95. Patel, H.R., A.M. Shekh, G.C. Pate, D.S. Mistry and C.T. Patel. 2002. Yield and quality of potato in relation to different dates of planting. *J. of Indian Potato Assoc.* 27 : 87–90.

96. Patel, R.N., N.H. Patel, S.K. Pandey, V.H. Kanbi and C.K. Patel. 2006. Yield performance and shelf life of some advanced potato hybrids in Gujarat. *Potato J.* 33 (3–4): 149–150.

97. Paul Vijay, R. Ezekiel and G.S. Shekhawat. 2002. Traditional method of potato storage in changing scenario. *Indian Farming*. August. 11–13 & 18–19.

98. Paul, V. and R. Ezekiel. 2005. Changes in temperature and relative humidity in heap and pit during storage of potatoes. *Potato Journal*. 32 (3- 4). 205–206.

99. Purohit, A.N., Mukhtar Singh, V.N. Banerjee, R.S. Raghow, Kshetrapal and A.K. Singh. 1970. Physiological studies on Indian Potato Varieties.1. Regional and seasonal adoptability, haulam growth, tuber initiation and number. *Indian J. Pl. Physiology* 13 : 240–248.

100. Pushkarnath and K.K. Nirula. 1970. Aphid warning for production of seed potato in subtropical of India. *Indian J. Agric. Sci.* 40 : 1061–1071.

101. Roy, K. and V.C. Srivastwa. 1996. Effect of seed size and seed rate on yield and economics of potato Cv. Kufri Bahar. *J. Indian Potato Assoc.* 23 (1&2): 85–87.

102. Roy, K. and V.P. Jaiswal. 1997. Effect of physiological age of seed tuber on sprout growth and yield of early-planted potato. *Indian J. Agric. Sci.* 67(4): 159–160.

103. Saini, S.S. 1978. Potash checks frost damage in potato. *J. Indian Potato Assoc.* 5: 108–109.

104. Sale, P.J.M. 1979. Growth of potatoes (*Solanum Tuberosum* L.) to the small tuber stages at related to soil temperature. *Aust. J. of Agric. Res.* 30: 667–675.

105. Sannigrahi, A.K. and B.C. Borah. 2002. Influence of black polythene and organic mulches on tomato and okra production in Assam. *Veg. Sci.* 29: 92–93.

106. Schipper, P.A. 1977. The rate of respiration of potato tubers during storage. III. Relationship between rate of respiration, weight loss and other variables. *Potato Res.* 20: 173–88.

107. Shekhawat, G.S., B.P. Singh and M.D. Jeswani. 1993. Soil and tuber borne diseases of potato. Tech. Bull. No. 41, CPRI, Shimla: 47.

108. Singh, A.N. 1990. In: "Commercial adoption of TPS technology – prospects and problems" (P.C. Gaur, Ed.) CPRI, Shimla: 49–53.

109. Singh Azad., B.K. Nehra, S.C. Khurana and Narendra Singh. 1997. Influence of plant density and fertility level on growth and yield in seed crop on potato. *J. Indian Potato Assoc.* 24 (1 – 2): 17–23.

110. Singh, A.N. and Janardan Jee. 1990. Economics of potato production from TPS and response of farmers in Bihar. Commercial Adoption of TPS technology – Prospects and Problems. (P.C. Gaur, Ed) CPRI, Shimle: 94–97.

111. Singh, A., B.K. Nehra, S.C. Khurana and N. Singh. 1997. Influence of plant density and geometry on growth and yield in seed crop of potato. *J. Indian Potato Assoc.* 24 (1–2): 24–30.

112. Singh, B and R. Ezekiel. 2005. Isopropyl N–(3–chlorophenyl) carbamate (CIPC) residues in potatoes treated with fog and dust formulations during storage. *Potato Journal.* 32 (3- 4): 213–214.

113. Singh, B and R. Ezekiel. 2003. Economics of potato storage at 10–12^{0}C with CIPC treatment in commercial cold stores. *J. Indian Potato Assoc.*30 (1–2): 171–172.

114. Singh, B. and R. Ezekiel. 2003. Processing quality of potatoes stored at 10–12^0 C with CIPC treatment in commercial cold stores. In: Abstract, Conference on processing and export potential of potatoes within Asia. March 10, 2003, IPA, CPRI and CIP, Modipuram, Meerut: 35–36.

115. Singh, B., H.N. Kaul and R. Ezekiel. 2004. Effect of CIPC dusting on potatoes during non – refrigerated storage: Sprout suppression and residues. *J. Food Science and Tech.* 41: 550–553.

116. Singh, J.P., A.Mehta and H.N. Kaul. 1995. A simple scheme to measure respiration rate of potato tuber with CO_2 gas analyser. *J. Indian Potato Assoc.* 22(1&2): 70-73

117. Singh, Kirti and R.P. Gangwar. 1972. Effect of mulching in the production of potato (*Solanum tuberosum* L.). *Haryana J. of Hort. Sci.* 1: 39–45.

118. Singh L., and S.N.S. Srivastava. 1983. Variation in temperature and humidity in a farm store and potato losses. *J. Indian Potato Assoc.* 10: 140–144.

119. Singh, M. 1983. In: Potato production, Storage and Utilization. (B.B. Nagaich, Ed.), Indian Potato Association. Shimla: 241-42.

120. Singh, N., and Braham Singh. 2003. Vegetable production in Ladakh. Field Research Laboratory, Leh – Ladakh (DRDO).

121. Singh, N, S.K. Dwivedi and Eli Paljor. 1999. Protected vegetable cultivation in Ladakh. Field Research Laboratory (DRDO),Leh.

122. Singh, N., B. Raut, N. Kumar and JN Mohanty. 2003. Off season potato production in trench in clod arid conditions of Leh – Ladakh. *J. Indian Potato Assoc.* 30 (1–2): 91–92.

123. Singh, N and B. Singh. 2003. Effect of mulching on different vegetables. Vegetable production in Ladakh. Field Research Laboratory, Defence Research and Development Organisation, Leh – Ladakh (J & K): 98–99.

124. Singh, N. 1994. Yield response of potato Cv. Kufri Swarna to varying seed size and spacing. *J. Indian Potato Assoc.* 21: 231–233.

125. Singh, N. 1997. Effect of seed size and seed rate on yield and economics of potato under rainfed conditions of Nilgiris hills. *J. Indian Potato Assoc.* 24 (3 & 4): 169–170.

126. Singh, N., Y.S. Malik, M.L. Pandita and B.K. Nahra. 1993. Effect of seed size, spacing and haulm cutting dates on yield of seed size tubers in potato Cv. Kufri Badshah. *Haryana J. Hort. Sci.* 22: 311–315.

127. Singh, N. and B. Singh. (2003). Analysis of Ladakh's soils. Vegetbale production in Ladakh. Field Research Laboratory, Defence Research and Development Organisation, Leh, Ladakh (J & K): 17–26.

128. Singh P.H., V.K. Chandla, S.M. Paul Khurana, I.D. Garg, K.C. Thakur, S. Singh and S.S. Lal. 2003. Potato cultivation in north eastern India. Technical Bulletin No. 34 (E). Central Potato Research Institute, Shimla.

129. Singh, S., V.K. Garg, S. Kumar and G.S. Shekhawat. 2000. Potato diseases, pest and their identification. Seed Potato Production Manual. CPRI – Shimla: 58–74.

130. Singh, S., V.K. Garg, S. Kumar and G.S. Shekhawat. 2000: Basic requirement for seed production. Seed Potato Production Manual, CPRI, Shimla: 5.

131. Singh, S.V., S.K. Pandey, D. Kumar, P. Kumar and S.M. Paul Khurana. 2002. Production of potatoes for processing. Extention Bulletin No. 33. CPRI, Shimla: 3.

132. Singh, J., A.N. Singh and P.C. Pandey. 1992. True potato seed for potato production in India. Technical Bulletin No. 24. Central Potato Research Institute, Shimla.

133. Struve, D.K. and B.C. Moser. 1984. Auxin effect on root regeneration of scarlet oak seedlings. *Journal American Society Horticulture Science* 109: 91–95.

134. Suman., Y.S. Malik and S.C. Khurana. 2003. Effect of fertilizer, spacing and crop duration on growth and yield of potato. *J. Indian Potato Assoc.* 30 (1–2): 87–88.

135. Thakur, K.C., P.C. Pandey and S.K. Pandey. 2003. Producing potato through TPS. *Indian Horticulture.* April June: 32–34 .

136. Trehan, S.P. and R.C. Sharma. 1999. Micronutrient requirement of potato. Technical Bulletin No.23 (Revised). CPRI,Shimla.

137. Upadhya, M.D. 1983. Studies on potato reproductive biology for TPS research in the year 2000. In Research for the potato in the year 2002. W.J. Hooker (ed.). CIP, Lima, Peru: 180–183.

138. Upadhya, M.D., K.C. Thakur, A. Juneja and M.S. Kadian. 1984. True potato seed production: Flowering, quality and economics. An innovative method for propagating potatoes, CIP, Lima, Peru: 117–147.

139. Upadhya, M.D., R. Chandra and Y.K. Sharma. 1981. Potato embryo characteristics and Variation. *True Potato Seed Letter*, 2:1.

140. Upadhya, M.D., T.R. Dayal, R. Chandra and A.N. Singh. 1986. Potential of true potato seed (TPS) for producing commercial potato crop in India. *J. Indian Potato Assco.* 13: 29–34.

141. Upadhya, M.D., M.S. Kadian, K.C. Thakur and V.S. Khatana. 1990. Potential of true potato seed (TPS) for potato production in different agro-climatic zones of India. (Abstract). *European Assoc Potato Research.* II Triennial Conference, Edinburgh, U.K, 8–13 July.

142. Upadhya, N.C. and R.C. Sharma. 1987. Varietal differences in fertilizer use efficiency of N in potato. *Journal of the Indian Society of Soil Science* 35: 654–660.

143. Varma, R.B. and R.D. Singh. 1996. Performance of TPS progenies for seedling tuber production in Nursery Beds. *J. Indian Potato Assoc.* 23 (1 & 2): 59–61.

144. Vaghela, K.O., N.H. Patel, R.N. Patel and C.K. Patel. 2007. Performance of parents for true potato seed attributes at Deesa (Gujrat). *Potato J.* 34 (1–2): 53–54.

145. Verma, S.S., J.R. Sharma and K.C. Joshi. 1973. Sugar content of potato tuber. *Indian J. Agric. Sci.* 43: 743–745.

146. Verma, S.C., T.R. Sharma and S.M. Verma. 1974. Effect of extended high temperature storage on weight losses and sugar content of potato tubers. *Indian J. Agric. Sci.* 44: 702–706.

147. Wadhwa B.K., M.L. Pandita and S.C. Khurana. 2000: Effect of spacing on ware crop of potato raised from seedling. *J. Indian Potato Assoc.* 27 (1–2): 45- 46.

148. Wilson, A.M., A.A. Bushway and R.J. Bushway. 1981. Residue analysis of isopropyl N–(3–chlorophenyl) carbamate in fruits and vegetables using high–performance liquid chromatography. *J. Agri and Food Chem.* 29: 746–749.

149. Wurr, D.C.E., S.J.R. Fellow and E.J. Allen. 1992. Determination of optimum tuber plant density in potato verity Penltard Squira, Cara, Estima, Maris Piper and King Edward. *J. Agric. Sci.* 119: 35–44.

150. Yamaguchi, M., Timm and A.R. Spuur. 1964. Effect of soil temperature on growth and nutrition of potato plants and tuberization, composition and periderm structures of tuber. *Proc. Am. Soc. of Hort. Sci.* 84: 412–423.

151. Yamaguchi, T. and H.E. Street. 1977. Stimulation of the growth of excised cultured root of soyabean by abscisic acid. *Annals of Botany* 41: 1129–1133.

❑❑❑